BRIDGE DECK ANALYSIS

SECOND EDITION

Eugene J. OBrien and Damien L. Keogh
Department of Civil Engineering
University College Dublin, Ireland

Alan J. O'Connor
Trinity College Dublin, Ireland

Chapter 4 written in collaboration with the authors by
Barry M. Lehane
Department of Civil, Environmental and Mining Engineering
University of Western Australia

CRC Press
Taylor & Francis Group
Boca Raton London New York

CRC Press is an imprint of the
Taylor & Francis Group, an **informa** business

BRIDGE DECK ANALYSIS

SECOND EDITION

Eugene J. OBrien and Damien L. Keogh
Department of Civil Engineering
University College Dublin, Ireland

Alan J. O'Connor
Trinity College Dublin, Ireland

Chapter 4 written in collaboration with the authors by
Barry M. Lehane
Department of Civil, Structural and Environmental Engineering
Trinity College Dublin, Ireland

CRC Press
Taylor & Francis Group
Boca Raton London New York

CRC Press is an imprint of the
Taylor & Francis Group, an **informa** business

CRC Press
Taylor & Francis Group
6000 Broken Sound Parkway NW, Suite 300
Boca Raton, FL 33487-2742

First issued in paperback 2019

© 2015 by Taylor & Francis Group, LLC
CRC Press is an imprint of Taylor & Francis Group, an Informa business

No claim to original U.S. Government works

ISBN-13: 978-1-4822-2723-9 (hbk)
ISBN-13: 978-0-367-86939-7 (pbk)

Library of Congress Cataloging-in-Publication Data

OBrien, Eugene J., 1958-
   Bridge deck analysis / Eugene OBrien, Damien Keogh, Alan O'Connor. -- Second edition.
      pages cm
   Includes bibliographical references and index.
   ISBN 978-1-4822-2723-9 (hardback)
   1. Bridges--Floors. 2. Structural analysis (Engineering) I. Keogh, Damien L., 1969- II. O'Connor, Alan (Bridge engineer) III. Title.

TG325.6.O27 2014
624.2'83--dc23
                                                     2014028237

Visit the Taylor & Francis Web site at
http://www.taylorandfrancis.com

and the CRC Press Web site at
http://www.crcpress.com

This book is dedicated to Sheena, Margaret and Mette.
Thank you for your endless patience.

Contents

Preface

This edition arose from a suggestion by Alan O'Connor that our book should include chapters on reliability theory. However, when we took a closer look, we found that the entire book was in need of a major update; and what started as minor revisions became a big undertaking. New research has changed the way that soil/structure interaction is treated in Chapter 4. We decided to drop the text on moment distribution in Chapter 3 and added a new section to give examples of how to analyse for the effects of creep. A lot has changed over the years. Grillage analysis is surely declining in popularity as plate finite-element (FE) programs are widely available, and most engineers are now familiar with the basics of FE theory. We have retained grillage analysis for now, but we de-emphasise it and have greatly expanded the sections on 3-D brick finite elements. The old references to the British Standard BS5400 are now gone, and the text is consistent with the Eurocodes and AASHTO standards. We have kept with our tradition of taking the reader through big examples in considerable detail. The feedback we get is that young engineers find this really useful.

In many ways, we have grown up with this book. Damien Keogh was just a graduate when we wrote the first edition, and he is now a project engineer with the international firm of consultants, Rambøll. Eugene OBrien was a junior lecturer when he was working on the first edition, and he is now a professor and a company director at Roughan O'Donovan Innovative Solutions. It has been a pleasure to update the book to reflect the many changes that have happened since the 1990s. We hope that the readers will agree that it has been worthwhile.

Acknowledgements

Several people helped us in the preparation of the second edition. Dr. Donya Hajializadeh, in particular, invested a great deal of time in running the analyses for examples in Chapters 3, 4 and 7, and she was particularly patient when the numbers changed and re-analysis was required. Rachel Harney, Cathal Leahy and Jennifer Keenahan also contributed analysis and figures essential to the explanation of complex concepts. On technical issues, Marcos Sanchez Sanchez was an immense resource; he is an outstanding bridge engineer, and he gave most generously of his time. Aonghus O'Keeffe, Arturo González, Bernard Enright, Colin Caprani and Cathal Leahy were also most helpful on technical questions. The Ministry of Infrastructure and the Environment of the Netherlands, Rijkswaterstaat, is acknowledged for making available weigh-in-motion data. Dr. Ib Enevoldsen of Rambøll Consulting Engineers is specifically thanked for his contribution and for making available the examples in Chapter 9, which form such an important part of the new edition. Finally, we would like to thank Dr. Arturo González for taking on extra lectures and Dr. Atorod Azizinamini with Florida International University for hosting a sabbatical in Miami, where much of the work for the second edition was completed. The cover photograph is provided courtesy of Roughan and O'Donovan Consulting Engineers.

Disclaimer

This publication presents many advanced techniques, some of which are novel and have not been exposed to the rigours of time. The material represents the opinions of the authors and should be treated as such. Readers should use their own judgement as to the validity of the information and its applicability to particular situations and check the references before relying on them. Sound engineering judgement should be the final arbiter in all stages of the design process. Despite the best efforts of all concerned, typographical or editorial errors may occur, and readers are encouraged to bring errors of substance to our attention. The publisher and authors disclaim any liability, in whole or in part, arising from information contained in this publication.

Authors

Dr. Eugene OBrien is professor of civil engineering at the University College Dublin (UCD), Ireland. After completing his PhD, Dr. OBrien worked for 5 years in the industry before becoming a lecturer in 1990 at Trinity College Dublin. Since 1998, he has been a professor of civil engineering at UCD. He has personally supervised 26 PhDs to completion and has published 220 technical papers and one other book. He has a significant track record of participation in European framework projects since the mid-1990s and, at the time of writing, leads a €2 million national project, *PhD in Sustainable Development*, funded by the Irish Research Council. He is also the UCD Principal Investigator on *Next Generation Bridge Weigh-in-Motion*, a $1 million project funded jointly by the Science Foundation Ireland, Invest Northern Ireland and the American National Science Foundation (NSF). As well as his academic work, Dr. OBrien is involved in the commercialisation of research as the director of Roughan O'Donovan Innovative Solutions. In that role, he leads the FP7 projects, *Long Life Bridges* and *InfraRisk*, and is a partner in the Research for SME project, *BridgeMon*.

Dr. Damien Keogh, BSc Eng, PhD, is a senior bridge design engineer and project manager in the International Bridges Department with Rambøll in Copenhagen, Denmark. He is a chartered engineer and member of the Institution of Engineers of Ireland. He has extensive international experience in bridge design and project management, having worked in Ireland, the Middle East, India and Denmark. His experience varies from single-span precast concrete road bridges up to large composite steel and concrete cable stayed bridges. At the time of writing, he is working on the Queensferry Crossing: a new 2.7 km road bridge across the Firth of Forth in Scotland where Rambøll are the lead designers.

Prof. Alan O'Connor, BA, BAI, PhD, is an associate professor in the Department of Civil Engineering at Trinity College Dublin, Ireland. He is a chartered engineer and a fellow of the Institution of Engineers of Ireland. He has extensive national/international experience in infrastructural risk analysis and probabilistic safety assessment. He has advised clients such as Irish Rail, The Irish National Roads Authority, The Danish Roads Directorate, Danish Railways, Swedish Railways, The Norwegian Roads Authority and the Ministry of Infrastructure and the Environment of the Netherlands. At Trinity College Dublin, the research group that he leads is focused on investigating infrastructural asset management and optimised whole life management, cross asset maintenance optimisation, structural health monitoring, stochastic modelling of engineering systems, risk analysis of critical infrastructure for extreme weather events and structural reliability analysis.

Chapter 1

Introduction

1.1 INTRODUCTION

A number of terms are illustrated in Figure 1.1, which are commonly used in bridge engineering.

As shown in Figure 1.1, all parts of the bridge over the bearings are referred to as super-structure, whereas parts below the bearings are referred to as substructures. The main body of the bridge superstructure is known as the deck and can consist of a main part and can-tilevers, as illustrated. The longitudinal direction is defined as the direction of span, and transverse is the direction perpendicular to it.

There may be upstands or downstands at the ends of the cantilever for aesthetic purposes, and there may be a parapet to retain the vehicles on the bridge. Bridge decks are frequently supported on bearings, which transmit the loads to abutments at the ends or to piers or walls elsewhere. Joints may be present to facilitate expansion or contraction of the deck.

1.2 FACTORS AFFECTING STRUCTURAL FORM

In recent decades, it has been established that a significant portion of the world's bridges are not performing as they should. In some cases, bridges are carrying significantly more traffic load than originally intended. However, in many others, the problem is one of durability – the widespread use of de-icing salt on roads has resulted in the ingress of chlorides into concrete. This is often associated with joints that are leaking or with details that have resulted in chloride-contaminated water dripping onto substructures. Problems have also been reported with post-tensioned concrete bridges in which inadequate grouting of the ducts has led to corrosion of the tendons.

The new awareness of the need to design durable bridges has led to changes of atti-tudes towards bridge design. There is now a significant move away from bridges that are easy to design towards bridges that will require less maintenance. The bridges that were easy to design were usually determinate, for example, simply supported spans and cantile-vers. However, such structural forms have many joints that are prone to leakage and have many bearings that require replacement over the lifetime of the bridge. The modern trend is towards bridges that are highly indeterminate and that have few joints or bearings.

The structural forms of bridges are closely interlinked with the methods of construction. The methods of construction, in turn, are often dictated by particular conditions on site. For example, when a bridge is to be located over an inaccessible place, such as a railway yard or a deep valley, the construction must be carried out without support from below. This immediately limits the structural forms to those that can be constructed in this way.

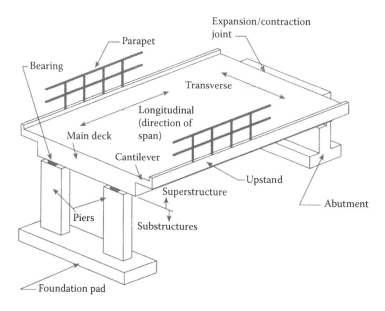

Figure 1.1 Portion of bridge illustrating bridge engineering terms.

The method of construction also influences the distributions of moment and force in a bridge. For example, in some bridges, steel beams carry the self-weight of the deck, whereas composite steel and in situ concrete carry the imposed traffic loading. Various alternative structural bridge forms and methods of construction are presented in the following sections.

1.3 CROSS SECTIONS

1.3.1 Solid rectangular

The solid rectangular section, illustrated in Figure 1.2a, is not a very efficient structural form, as the second moment of area of a rectangle is small relative to its area (and weight). Such a bridge is generally constructed of reinforced concrete, particularly for the shorter spans, or prestressed concrete. Due to the inefficiency of this structural form, the stresses induced by the self-weight of the concrete can become excessive. However, the shuttering

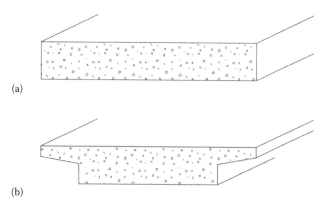

Figure 1.2 In situ solid rectangular section: (a) without cantilevers; (b) with cantilevers.

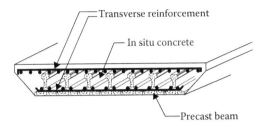

Figure 1.3 Precast and in situ solid rectangular section.

costs for a bridge with a flat soffit are relatively low, and the reinforcement is generally simple. As a result, this form of cross section is often the most cost-effective for shorter spans (up to about 20 m). As can be seen in Figure 1.2, rectangular bridges can be constructed with or without cantilevers.

Comparing bridges of the same width, such as illustrated in Figure 1.2a and b, it can be seen that the bridge with cantilevers has less weight, without much reduction in the second moment of area. However, what is often the more important advantage of cantilevers is the aesthetic one, discussed in Section 1.8. Solid rectangular sections can be constructed simply from in situ concrete as illustrated in Figure 1.2. Such construction is clearly more economical when support from below the bridge is readily available. When this is not the case, for example, over railway lines or deep waterways, a rectangular section can be constructed using precast, pre-tensioned, inverted T-sections, as illustrated in Figure 1.3. Holes are cast at frequent intervals along the length of such beams to facilitate the threading-through of transverse bottom reinforcement. In situ reinforced concrete is then poured over the precast beams to form the complete section. With this form of construction, the precast beams must be designed to carry their self-weight plus the weight of the (initially wet) in situ concrete. The complete rectangular section is available to carry other loading.

1.3.2 Voided rectangular

For spans in excess of approximately 20 m, solid rectangular sections become increasingly less cost-effective due to their low second-moment-of-area-to-weight ratio. For the span range of 20–30 m, it is common practice in some countries to use in situ concrete with polystyrene 'voids', as illustrated in Figure 1.4. These decks can be constructed from ordinary reinforced concrete or can be post-tensioned. Including voids in a bridge deck increases the cost for a given structural depth because it adds to the complexity of the reinforcement, particularly reinforcement for transverse bending. However, it reduces considerably the self-weight and the area of concrete to be prestressed without significantly affecting the second moment of area. The shuttering costs are also less than for in situ concrete T-sections, which are described below. Hence it is, in some cases, the preferred solution, particularly when the designer wishes to minimise the structural depth.

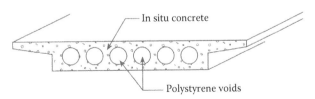

Figure 1.4 Voided slab section with cantilevers.

It is essential in such construction to ensure that sufficient stays are provided to keep the voids in place when the concrete is poured and to prevent uplift due to flotation. This problem is not so much one of steel straps failing as of grooves being cut in the polystyrene by the straps. Concerns have been expressed about voided slab construction over the lack of inspectability of the concrete on the inside of the void, and there are many countries where this form is virtually unknown.

It is common practice to treat voided slabs as solid slabs for the purposes of analysis, provided that the void diameter is less than 60% of the total depth. Regardless of the diameter-to-depth ratio, the voids must be accounted for when considering the design to resist transverse bending. Guidance is given on the analysis of this type of deck in Chapter 6.

1.3.3 T-section

The T-section is commonly used for spans in the range of 20–40 m as an alternative to voided slab construction. The T-section is a less efficient structural form as it tends to have more material close to the centroid of the bridge than a voided slab. As a result, the section tends to be deeper for a given span. In situ T-section decks, illustrated in Figure 1.5, are

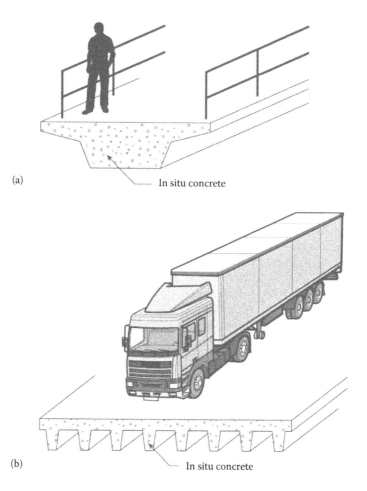

(a)   In situ concrete

(b)   In situ concrete

Figure 1.5 In situ concrete T-sections: (a) single web such as might be used for a pedestrian bridge; (b) multiple webs such as would be used for wider decks.

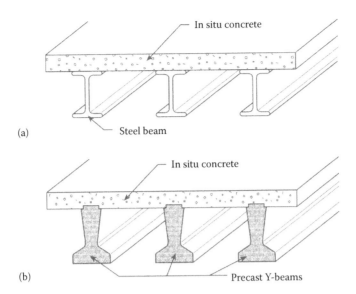

Figure 1.6 T-sections: (a) composite steel and concrete; (b) composite precast Y-beam and in situ concrete.

more expensive in terms of shuttering costs than voided slabs but have a major advantage in that all of the bridge deck is totally inspectable.

Precast concrete or steel forms of T-section (Figure 1.6) are often favoured because of cost, durability or sometimes a lack of access to support from below. These consist of pre-tensioned, prestressed concrete or steel beams placed in position along the length of the span. An in situ concrete slab, supported on permanent shuttering, spans transversely between the beams while acting as flanges to the beams longitudinally.

1.3.4 Box sections

For spans in excess of approximately 40 m, it becomes economical to use 'cellular' or 'box' sections, as illustrated in Figure 1.7. These have a higher second moment of area per unit

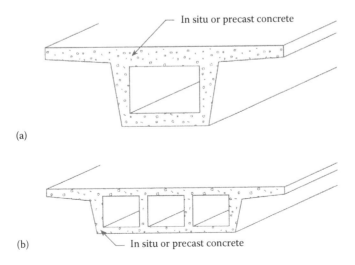

Figure 1.7 Box sections: (a) single cell; (b) multi-cellular.

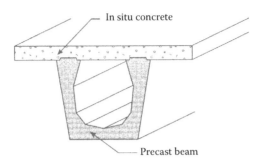

Figure 1.8 Composite precast and in situ box section.

weight than voided slab or T-sections. However, in their in situ form, they are only considered economical at higher spans. This happens when the structural depth reaches approximately 2 m, at which point it is easier for personnel to enter the void to recover the shuttering and, when the bridge is in service, to inspect the inside of the void. Box sections can also be constructed of precast concrete or can be composite with a precast, pre-tensioned U-section and an in situ concrete slab, as illustrated in Figure 1.8.

1.3.5 Older concepts

Many variations of the above structural forms have been used in the past and are evident in existing bridge stocks. For example, in the past, it was common practice to construct T-section decks using precast 'M-beams' (Figure 1.9). These have wider bottom flanges than the precast 'Y-beams' (Figure 1.6b) used more commonly today (Taylor et al. 1990). A disadvantage of the M-section is that it is difficult to compact the concrete properly at the top surface of the wide bottom flange. In the past, M-sections were often placed side by side with the bottom flanges within millimetres of each other. The analysis of this type of bridge is similar to that of any T-section bridge.

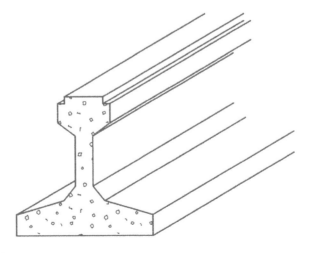

Figure 1.9 Precast M-beam.

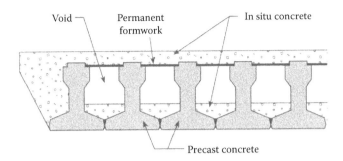

Figure 1.10 Pseudo-box section.

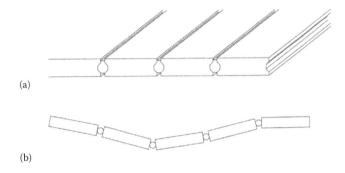

(a)

(b)

Figure 1.11 Shear-key deck: (a) section through small portion of deck; (b) assumed transverse deformation.

It was also common practice in the past to build bridges of 'pseudo-box' construction as illustrated in Figure 1.10. These were constructed of M-beams with in situ concrete near the bottom to form a void. The bottom in situ concrete was reinforced transversely by threading bars through holes cast in the M-beams. The section is more efficient than a T-section as more concrete is located away from the centroid. However, if water leaks into the voids, corrosion problems can result, and due to the nature of this structural form, assessment and repair are difficult. The structural behaviour of the pseudo-box section is similar to that of a small multi-cellular box section.

Another form of construction used in the past is the 'shear-key' deck, illustrated in Figure 1.11a. This consists of precast concrete slab strips joined using longitudinal strips of in situ concrete. The latter 'shear keys' are assumed to be capable of transferring shear force but not transverse bending moment as they have no transverse reinforcement. Thus, the transverse deformation is assumed to be as illustrated in Figure 1.11b; rotation is assumed to occur at the joints between precast units. Shear-key decks were popular for railway bridge construction in the United Kingdom, as the railway line could be reopened even before the in situ concrete was placed. However, they are no longer popular due to concerns about the durability of the in situ joints. It should be noted that the 'bulb-tee' girder bridge system, still popular in the United States, utilises a similar shear-key concept.

1.4 BRIDGE ELEVATIONS

The cross sections described above can be used in many different forms of bridge. Alternative bridge elevations and their methods of construction are described in the following sections.

1.4.1 Simply supported beam/slab

The simplest form of bridge is the single-span beam or slab, which is simply supported at its ends. This form, illustrated in Figure 1.12, is widely used when the bridge crosses a minor road or small river. In such cases, the span is relatively small, and multiple spans are infeasible and/or unnecessary. The simply supported bridge is relatively easy to analyse and construct but is disadvantaged by having bearings and joints at both ends. The cross section is often solid rectangular but can be of any of the forms presented above.

1.4.2 Series of simply supported beams/slabs

When a bridge crossing is too wide for an economical single span, it is possible to construct what is in effect a series of simply supported bridges, one after the other, as illustrated in Figure 1.13. Like single-span bridges, this form is relatively simple to analyse and construct. It is particularly favoured on poor soils where differential settlement of supports is anticipated – differential settlement generates no bending moment in this form of construction. A series of simply supported spans also has the advantage that, if constructed using in situ concrete, the concrete pours are moderately sized. In addition, there is less disruption to any traffic that may be below as only one span needs to be closed at any one time. However, there are a great many joints and bearings with the result that a series of simply supported beams/slabs is no longer favoured in practice.

A continuous beam/slab, as illustrated in Figure 1.14, has significantly fewer joints and bearings than simply supported spans. A further disadvantage of simply supported beams/slabs in comparison to continuous ones is that the maximum bending moment in the former is significantly greater than that in the latter. For example, the bending moment diagrams due to a uniformly distributed loading of intensity ω (kN/m) are illustrated in Figure 1.15. It can be seen that the maximum moment in the simply supported case is significantly greater (by 25%) than that in the continuous case. The implication of this is that the simply supported bridge deck needs to be correspondingly deeper.

1.4.3 Continuous beam/slab with full propping during construction

As stated above, continuous beam/slab construction has significant advantages over simply supported spans in that there are fewer joints and bearings, and the applied bending

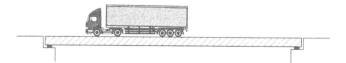

Figure 1.12 Simply supported beam or slab.

Figure 1.13 Series of simply supported beam/slabs.

Figure 1.14 Continuous beam or slab.

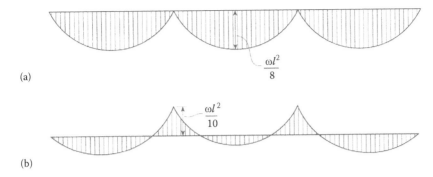

$$\frac{\omega l^2}{8}$$

(a)

$$\frac{\omega l^2}{10}$$

(b)

Figure 1.15 Bending moment diagrams due to uniform loading of intensity ω: (a) three simply supported spans of length, *l*; (b) one three-span continuous beam with span lengths, *l*.

moments are less. For bridges of moderate total length, the concrete can be poured in situ in one pour. This completely removes the need for any internal joints. However, as the total bridge length becomes large, the amount of concrete that needs to be cast in one pour can become excessive. This tends to increase cost as the construction becomes more of a batch process than a continuous one.

1.4.4 Partially continuous beam/slab

When support from below during construction is expensive or infeasible, it is possible to use precast concrete or steel beams to construct a partially continuous bridge. Precast concrete or steel beams are placed initially in a series of simply supported spans. In situ concrete is then used to make the finished bridge continuous over intermediate joints. Two forms of partially continuous bridge are possible. In the form illustrated in Figure 1.16, the in situ concrete is cast to the full depth of the bridge over all supports to form what is known as a diaphragm beam. Elsewhere, the cross section is similar to that illustrated in Figure 1.6.

In the alternative form of a partially continuous bridge, illustrated in Figure 1.17, continuity over intermediate supports is provided only by the slab. Thus, the in situ slab alone is required to resist the complete hogging moment at the intermediate supports. This is possible due to the fact that members of low structural stiffness (second moment of area) tend to attract low bending moment. The slab at the support in this form of construction is particularly flexible and tends to attract a relatively low bending moment. There is concern among some designers about the integrity of such a joint, as it must undergo significant rotation during the service life of the bridge. Further, as the main bridge beams rotate at their ends, the joint must move longitudinally to accommodate this rotation, as illustrated in Figure 1.18. However, there are many bridges of this form in service around the world, and they appear to be performing well.

In partially continuous bridges, the precast concrete or steel beams carry all the self-weight of the bridge. For a two-span bridge, for example, this generates the bending moment diagram illustrated in Figure 1.19a. By the time the imposed traffic loading is applied, the bridge is continuous, and the resulting bending moment diagram is as illustrated in Figure 1.19b. The total bending moment diagram will be a combination of that due to self-weight and other loading. Unfortunately, due to creep, self-weight continues to cause deformation in the bridge after it has been made continuous. At this stage, it is resisted by a continuous rather than a simply supported beam/slab, and it generates a distribution of bending moment more like that of Figure 1.19b than Figure 1.19a. This introduces a complexity into

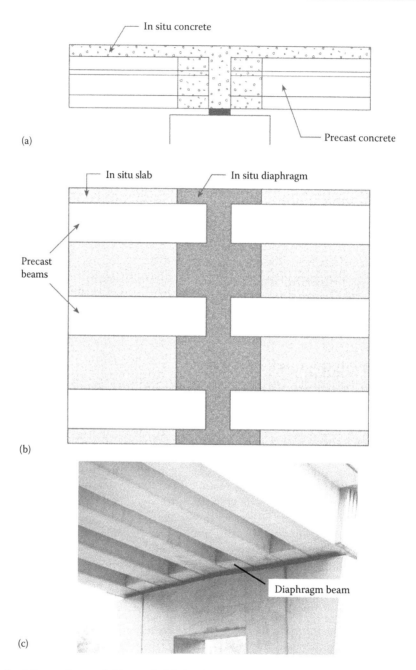

Figure 1.16 Partially continuous bridge with full-depth diaphragm at intermediate supports: (a) elevation; (b) plan view from below; (c) picture of a typical diaphragm beam.

the analysis, compounded by a difficulty in making accurate predictions of creep effects. The issue is discussed further in Section 3.7.

The great advantage of a partially continuous construction is in the removal of all inter-mediate joints while satisfying the requirement of construction without support from below. The method is also of a continuous rather than a batch form as the precast beams

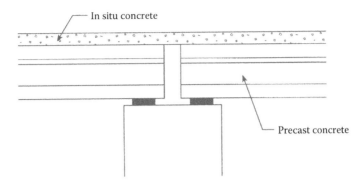

Figure 1.17 Partially continuous bridge with continuity provided only by the slab at intermediate supports.

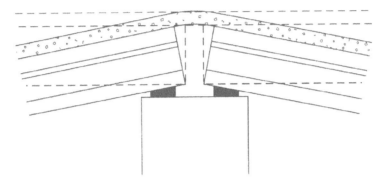

Figure 1.18 Joint rotation at intermediate support of a partially continuous bridge of the type illustrated in Figure 1.17.

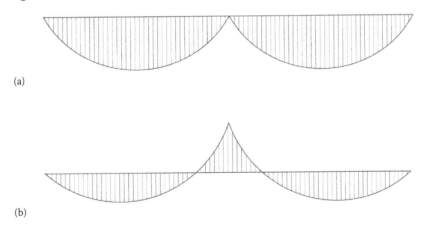

Figure 1.19 Typical distribution of bending moment in a two-span, partially continuous bridge: (a) bending moment due to self-weight; (b) bending moment due to loading applied after bridge has been made continuous.

can be constructed at a steady pace, starting even before work has commenced on site. Construction on site is fast, resulting in minimum disruption to any existing traffic passing under the bridge.

A significant disadvantage is that, while intermediate joints have been removed, intermediate bearings are still present with their associated maintenance implications. Particularly for the form illustrated in Figure 1.17, two bearings are necessary at each intermediate support.

1.4.5 Continuous beam/slab: Span-by-span construction

For the construction of particularly long bridges when access from below is expensive or infeasible, in situ construction, one span at a time, can be a viable option. This is achieved using temporary formwork supported on the bridge piers as illustrated in Figure 1.20a.

Proprietary post-tensioning couplers, such as illustrated in Figure 1.21, can be used to achieve continuity of prestressing across construction joints. In this form of construction, the joint where one concrete pour meets the next is designed to transmit bending moment

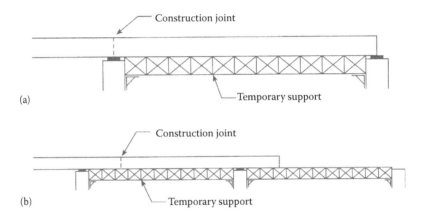

Figure 1.20 Temporary support system for span-by-span construction: (a) construction joint over intermediate pier; (b) construction joint at quarter span.

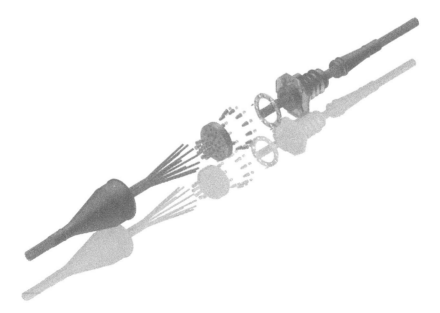

Figure 1.21 Exploded view of post-tensioning coupler to transmit prestress forces across a construction joint. (Courtesy of CCL.)

and shear force, and is not intended to accommodate movements due to thermal or creep/shrinkage effects. The joint may sometimes be located at the quarter-span position, as illustrated in Figure 1.20b, where bending moments and shear forces are relatively small.

In particularly long continuous beams/slabs, an intermediate joint may become necessary to relieve stresses due to expansion/contraction. It has been said that joints should be provided every 100 m at least. However, this figure is constantly being revised upwards as the problems of bridge joints in service receive ever more attention.

1.4.6 Continuous beam/slab: Balanced cantilever construction

When the area under a long bridge is difficult to access, it is often economical to construct bridges by the balanced cantilever method. Expensive access means a lesser number of longer spans, which rules out full-span precast beams. The cross section is generally of the box type constructed either of in situ concrete or precast segments of relatively short length (4–5 m longitudinally). This form of bridge is generally made of post-tensioned prestressed concrete.

The sequence of construction is illustrated in Figure 1.22. An intermediate pier is cast first and then a small part of the bridge deck known as the base segment (Figure 1.22a). This is prevented from rotation by a temporary prop or props connecting the deck to the foundation as illustrated. Such a prop is only capable of resisting a small out-of-balance loading, so it is necessary to have approximately equal lengths of cantilever on each side at all times during construction. Segments of deck are then added to the base segment, either alternately on opposing sides or simultaneously in pairs, one on each side. The segments are supported by a 'travelling form' connected to the existing bridge (Figure 1.22b) until such time as they can be permanently post-tensioned into place as illustrated in Figure 1.22c. Ducts are placed in all segments when they are first cast, in anticipation of the need to post-tension future segments at later stages of construction.

Segments can be cast in situ or precast; in the case of the latter, there is typically a 'shear key', as illustrated in Figure 1.22d, to provide a positive method of transferring shear between segments. Moment is transferred by the concrete in compression and by the post-tensioning tendons. While epoxy resin is commonly used to join segments, it does not normally serve any structural purpose.

Segments are added on alternate sides until they reach an abutment or another cantilever coming from the other side of the span. When cantilevers meet at mid-span, a 'stitching segment' is cast to make the bridge continuous as illustrated in Figure 1.23. Post-tensioning tendons are placed in the bottom flange and webs by means of 'blisters', illustrated in Figure 1.24, to resist the sagging moment that will exist in the finished structure due to applied traffic loading.

The bending moment in a balanced cantilever bridge is entirely hogging, while the bridge remains in the form of a cantilever. Thus, the moment due to self-weight during construction is such as illustrated in Figure 1.25a. After the casting of the stitching segment and completion of construction, the bridge forms a continuous beam, and the imposed service loading generates a distribution of moment such as that illustrated in Figure 1.25b. This form of bridge is quite inefficient as parts of it must be designed to resist a significant range of moments from large hogging to large sagging. Nevertheless, it is frequently the most economical alternative for construction over deep valleys when propping from below is expensive.

The analysis of balanced cantilever bridges is complicated by a creep effect similar to that for partially continuous beams. This is caused by a tendency for the distribution of moment due to self-weight to change in the long term from the form illustrated in Figure 1.25a towards the form illustrated in Figure 1.25b. This results from creep deformations that are still taking place after the bridge has been made continuous. Analysis for the effects of creep is discussed further in Section 3.7.

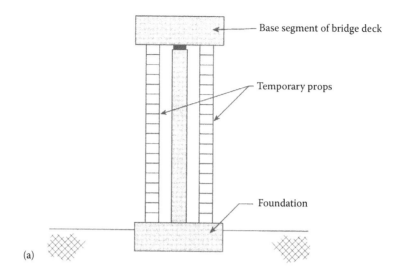

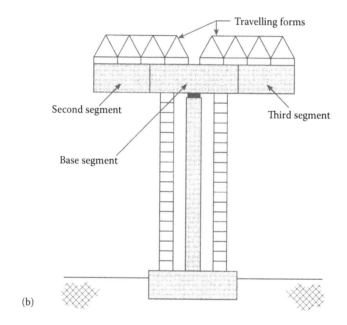

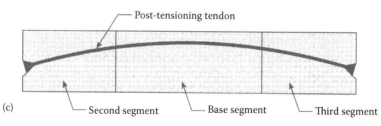

Figure 1.22 Balanced cantilever construction: (a) elevation of base segment and pier; (b) temporary support of segments; (c) sectional elevation showing tendon.

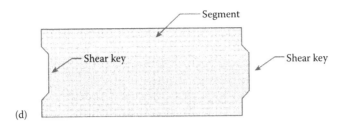

(d)

Figure 1.22 (Continued) Balanced cantilever construction: (d) precast segment.

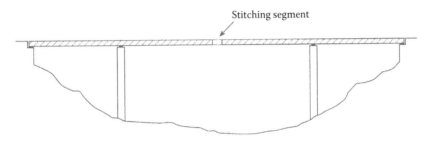

Figure 1.23 Casting of stitching segment.

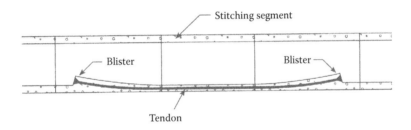

Figure 1.24 Blisters and tendon in the bottom flange (sectional elevation).

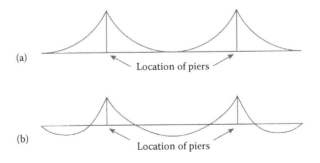

Figure 1.25 Distributions of bending moment in a balanced cantilever bridge: (a) due to self-weight during construction; (b) due to imposed loading after completion of construction.

1.4.7 Continuous beam/slab: Push-launch construction

For longer bridges, 'incremental-launch' or 'push-launch' becomes a viable alternative to balanced cantilever as a method of construction. In push-launch construction, a long segment is cast behind the bridge abutment as illustrated in Figure 1.26a. Hydraulic jacks are then used to 'push' this segment out into the first span to make way for the casting of another segment behind it (Figure 1.26b). This process is continued until the complete bridge has been constructed behind the abutment and pushed into place. When the deck is being pushed over intermediate supports, temporary sliding bearings are used to minimise friction forces.

The method has a considerable advantage of access. All of the bridge is constructed in the same place, which is easily accessible to the construction personnel and the plant. A significant disadvantage stems from the distribution of bending moment generated temporarily during construction. Parts of the deck must be designed for significant hog moment during construction as illustrated in Figure 1.27a. These same parts may be subjected to sag moment in the completed bridge as illustrated in Figure 1.27b. The effect is greater than in balanced cantilever construction as the cantilever length is the complete span length (as opposed to half the span length for the balanced cantilevers). This doubling of cantilever length has the effect of quadrupling the moment due to self-weight during construction. Bridges designed for push-launch construction, like those designed for balanced cantilever construction, must be designed for the creep effect and are subject to the associated complexity and uncertainty in design.

1.4.8 Arch bridges

For larger spans (in excess of approximately 50 m), the arch form is particularly effective. However, arches generate a significant horizontal thrust, as illustrated in Figure 1.28a, and

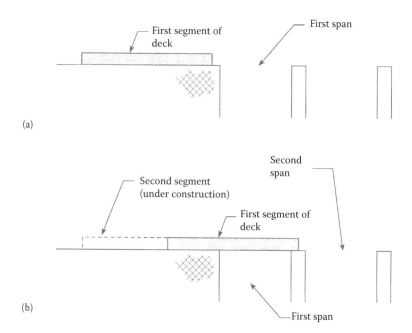

Figure 1.26 Push-launch construction: (a) casting of the first segment; (b) pushing of the partially constructed bridge over first span.

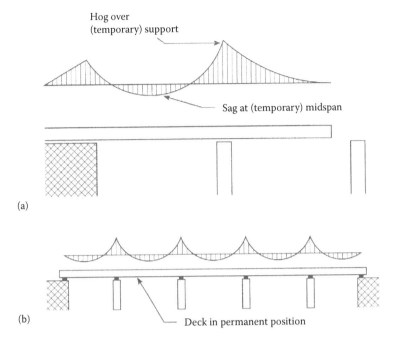

Hog over
(temporary) support

Sag at (temporary) midspan

(a)

(b)

Deck in permanent position

Figure 1.27 Distributions of bending moment in a push-launch bridge: (a) due to self-weight during construction; (b) due to imposed loading after completion of construction.

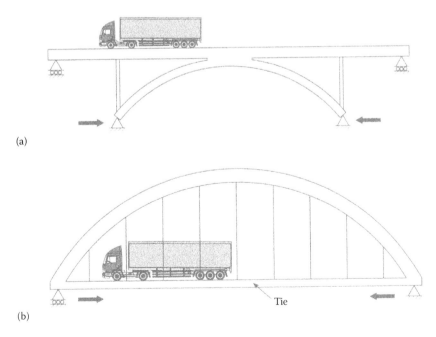

(a)

(b)

Tie

Figure 1.28 Arch bridges: (a) conventional form with deck over the arch; (b) tied arch with deck at base of arch.

are only a viable solution if it can be accommodated. Arches therefore work well if the bridge is located on a very sound foundation (such as rock). If this is not the case, an arch is still a possibility if it is tied as illustrated in Figure 1.28b. In a tied arch, the horizontal thrust is taken by the tie. Some engineers design bridges in an arch form for aesthetic reasons but articulate the bridge like a simply supported beam, as illustrated in Figure 1.29. This is perfectly feasible, but, as the bridge has no means to resist the horizontal thrust, it behaves structurally as a simply supported beam. While traditional masonry arches were designed to be completely in compression, modern concrete or steel arches have no such restriction and can be designed to resist bending as well as the axial compression generated by the arch form.

Concrete arches are particularly effective as concrete is very strong in compression. The arch action causes the self-weight to generate a compression, which has all the advantages of prestress but none of the disadvantages of cost or durability associated with tendons. Thus, the self-weight generates a distribution of stress, which is in fact beneficial and assists in the resistance of stresses due to imposed loading. Other advantages of arches are that they are aesthetically pleasing in the right environment, the structural depth can be very small and large clear spans can readily be accommodated. For example, while a continuous beam/slab crossing a 60-m motorway would normally be divided into two spans, an arch can readily span such a distance in one clear span, creating openness under the bridge that would not otherwise be possible. An additional major advantage is that arches require no bearings as it is possible to cast the deck integrally into the substructures. As can be seen in Figure 1.30, movements due to thermal expansion/contraction and creep/shrinkage do generate some stresses, but these are not as significant as those in the frame form of construction discussed below.

The principal disadvantage of concrete arches, other than the problem of accommodating the horizontal thrust, is the fact that the curved form requires curved shuttering, which is more expensive than would otherwise be the case. Also, if arches are located over inaccessible areas, considerable temporary propping is required to support the structure during construction.

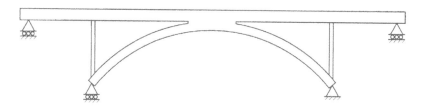

Figure 1.29 Simply supported beam bridge in the shape of an arch (note the roller support, which prevents horizontal thrust).

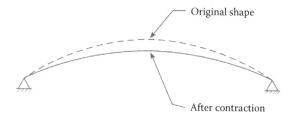

Figure 1.30 Deflected shape of arch subjected to thermal contraction.

1.4.9 Frame or box culvert (integral bridge)

Frame or box bridges, such as illustrated in Figure 1.31, are more effective at resisting applied vertical loading than simply supported beams/slabs. This is because, like the continuous beam/slab, the maximum bending moment tends to be less, as can be seen from the examples in Figure 1.32. However, accommodating movements due to temperature changes or creep/shrinkage can be a problem. The effects of deck shortening relative to the supports is to induce bending in the whole frame, as illustrated in Figure 1.33. If some of this shortening is due to creep or shrinkage, there is the usual complexity and uncertainty associated with such calculations. A further complexity in the analysis of frame bridges is that, unless the transverse width is relatively small, the structural behaviour is three-dimensional. Continuous slab bridges, on the other hand, can be analysed using two-dimensional models.

The minimal maintenance requirement of frame/box culvert bridges is their greatest advantage. There are no joints or bearings as the deck is integral with the piers and abutments. Given today's emphasis on maintenance and durability, this form of construction is very popular. Ever longer spans are being achieved. It is now considered that frame bridges of at least 100 m are possible. There are two implications for longer frame-type bridges, both relating to longitudinal movements. If the supports are fully fixed against translation, deck movements in such bridges will generate enormous stresses. This problem has been overcome by allowing the supports to slide, as illustrated in Figure 1.34. If the bridge is supported on piles, the axes of the piles are orientated so as to provide minimum resistance to longitudinal movement.

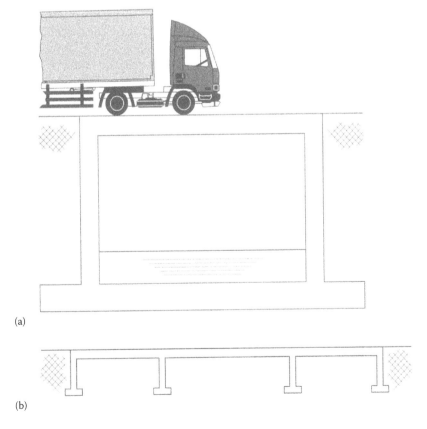

(a)

(b)

Figure 1.31 Frame/box culvert bridges: (a) box culvert; (b) three-span frame.

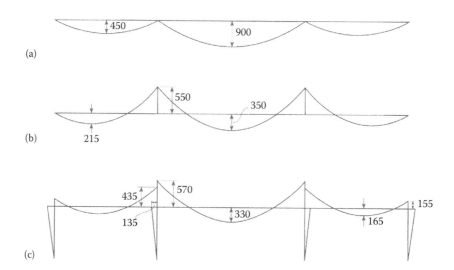

(a)

(b)   215

(c)

Figure 1.32 Typical distributions of bending moment: (a) simply supported spans; (b) continuous beam; (c) frame.

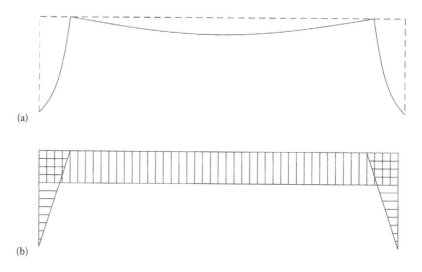

(a)

(b)

Figure 1.33 Effect of thermal contraction of deck in frame bridge: (a) deflected shape; (b) distribution of bending moment.

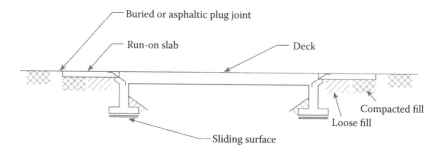

Figure 1.34 Sliding support and run-on slab in frame bridge.

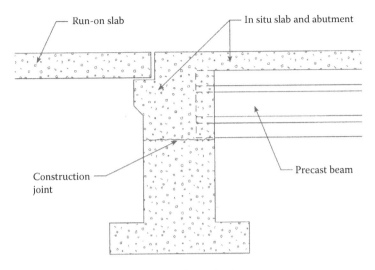

Figure 1.35 Composite precast and in situ concrete frame bridge.

The second implication of longer frame bridges is that the bridge moves relative to the surrounding ground. To overcome this, engineers specify 'run-on' slabs, as illustrated in Figure 1.34, which span over loose fill that is intended to allow the abutments to move. The run-on slab can rotate relative to the bridge deck, but there is no relative translation. Thus, at the ends of the run-on slabs, joints are required to facilitate translational movements. Such joints are remote from the main bridge structure and, if they do leak, will not lead to deterioration of the bridge itself.

A precast variation of the frame bridge is particularly popular. Precast, pre-tensioned concrete beams have a good record of durability and do not suffer from the problems associated with grouted post-tensioning tendons. These precast beams can be used in combination with in situ concrete to form a frame bridge as illustrated in Figure 1.35. Cross sections are typically of the form illustrated in Figure 1.6b. There are a number of variations of this form of construction, which are considered further in Chapter 4.

1.4.10 Beams/slabs with drop-in span

For ease of construction and of analysis, some older bridges were constructed of precast concrete with 'drop-in' spans. A typical example is illustrated in Figure 1.36. This bridge is determinate as the central 'drop-in' part is simply supported. The side spans are simply supported with cantilevers. The reactions from the drop-in span are transferred as loads to the ends of the cantilevers. The form has the disadvantage of having joints and bearings at the ends of the drop-in span as well as at the extremities of the bridge itself. However, it can readily be constructed over inaccessible areas. The drop-in span, in particular, can be placed

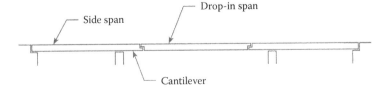

Figure 1.36 Beam bridge with drop-in span.

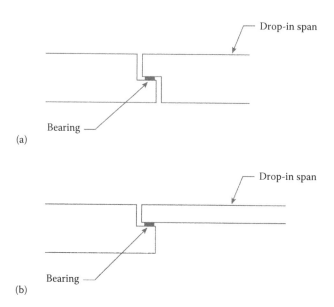

Figure 1.37 Halving joint at end of drop-in span: (a) traditional detail (no access); (b) alternative detail with access.

in position very quickly over a road or railway, requiring a minimum closure time. Thus, it is still popular in some countries for pedestrian bridges over roads.

In this form of construction, the joint and bearing detail at the ends of the drop-in span is particularly important. In older bridges of the type, two 'halving joints', as illustrated in Figure 1.37a, were used. This detail is particularly problematic as access to inspect or replace the bearings is extremely difficult. A more convenient alternative, which provides access, is illustrated in Figure 1.37b.

However, regardless of which alternative is chosen, halving joints frequently cause difficulty for a number of reasons:

- The joints tend to leak, which promotes corrosion of the halving joint reinforcement (see Figure 1.38).
- There are very high tensile and shear stresses at a point where the structural depth is relatively small.
- As shown in Figure 1.39, there can be difficulty finding space to provide sufficient reinforcement to resist all of the types of structural action that take place in the halving joint.

1.4.11 Cable-stayed bridges

Cable-stayed construction, illustrated in Figure 1.40, becomes feasible when the total bridge length is in excess of approximately 150 m and is particularly economical for lengths in the 200–400 m range. The maximum main span achievable is increasing all the time; the current limit is of the order of 1000 m. The concept of cable-stayed bridges is simple. The cables are only required to take tension, and they provide support to the deck at frequent intervals. The deck can then be designed as a continuous beam with spring supports. An analysis complication is introduced by a sag in the longer cables, which has the effect of

Figure 1.38 Rust staining indicates corrosion in a halving joint.

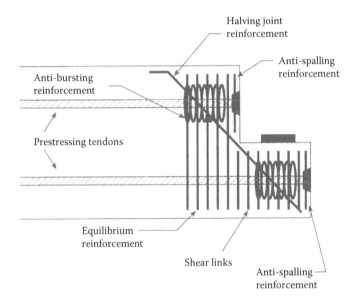

Figure 1.39 Reinforcement detail in halving joint.

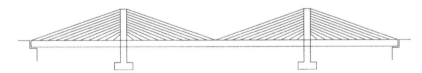

Figure 1.40 Cable-stayed bridge.

making the stiffness of the support they provide non-linear. It is also generally necessary to carry out a dynamic analysis for bridges of such slenderness. For spans of moderate length, the cross sections of cable-stayed bridges are often composite with steel beams and concrete slabs; for the longest spans, steel box section decks are used to reduce the bridge self-weight.

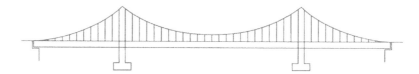

Figure 1.41 Suspension bridge.

The economy of the cable-stayed form stems from its ease of construction over inaccessible places. It readily lends itself to staged construction, with the cables being added as required to support successively placed segments of the deck. As for balanced cantilever bridges, segments are placed successively on alternate sides of the pylon. The towers in cable-stayed bridges tend to be relatively high over the deck: 25% of the main span is a common rule of thumb.

1.4.12 Suspension bridges

The longest bridges in the world, up to approximately 3000 m span,[1] are of the suspension type illustrated in Figure 1.41. In suspension bridges, the main cables are in catenary, and the deck hangs from them, applying a substantially uniform loading. It is a very elegant form of construction, and the towers tend to be much lower than the corresponding towers in cable-stayed bridges. However, suspension bridges are more expensive to construct than cable-stayed bridges as they are not particularly suited to staged construction, and the initial placing of the cables in position is onerous. Further, it is sometimes difficult to cater for the horizontal forces generated at the ends of the cables. The greatest disadvantage of suspension bridge construction is that it is extremely difficult to replace the cables if they become corroded. For these reasons, cable-stayed construction is generally favoured over suspension except for the longest spans.

1.5 ARTICULATION

Bridge design is often a compromise between the maintenance implications of providing joints and bearings and the reduction in stresses, which results from accommodating deck movements. While the present trend is to provide ever fewer joints and bearings, the problems of creep, shrinkage and thermal movement are still very real, and no one form of construction is the best for all situations.

The articulation of a bridge is the scheme for accommodating movements due to creep, shrinkage and thermal effects while keeping the structure stable. While this clearly does not apply to bridges without joints or bearings, it is a necessary consideration for those that do.

Horizontal forces are caused by braking and traction of vehicles, wind and accidental impact forces from errant vehicles. Thus, the bridge must have the capacity to resist some relatively small forces while accommodating movements.

In situ concrete bridges are generally supported on a finite number of bearings. The bearings usually allow free rotation but may or may not allow horizontal translation. They are generally of one of the following three types:

1. Fixed – no horizontal translation allowed
2. Free sliding – free to move horizontally in both directions
3. Guided sliding – free to move horizontally in one direction only

---

[1] At the time of writing, the Messina Straits bridge is being designed with a main span of 3300 m.

In many bridges, a combination of the three types of bearing is used. Two of the simplest forms of articulation are illustrated in Figure 1.42a and b where the arrows indicate the direction in which movements are allowed. For both bridges, A is a fixed bearing, allowing no horizontal movement. To make the structure stable in the horizontal plane, guided sliding bearings are provided at C and, in the case of the two-span bridge, also at E. These bearings are designed to resist horizontal forces such as the impact force due to an excessively high vehicle attempting to pass under the bridge. At the same time, they accommodate longitudinal movements such as those due to temperature changes. Free-sliding bearings are provided elsewhere to accommodate transverse movements. When bridges are not very wide (less than approximately 5 m), transverse movements are sometimes assumed to be negligible as illustrated in Figure 1.42c.

When bridges are not straight in plan, the orientation of movements tends to radiate outwards from the fixed bearing. This can be seen in the simple example illustrated in Figure 1.43a.

Creep, shrinkage or thermal movement results in a predominantly longitudinal effect, which causes AB to shorten by δ_1 to AB'. Similarly, BC shortens by δ_2 to BC'. However, as B has moved to B', C' must move a corresponding distance to C''. Overall, C moves a distance δ_1 parallel to AB and δ_2 parallel to BC. Combining these two vectors gives the direction of movement. If the strain is the same in AB and BC, these movements are proportional to the lengths, and the orientation of the vector CC'' is defined by the vector sum of $\overrightarrow{AB}$ and $\overrightarrow{BC}$, that is, $\overrightarrow{AC}$. Hence, the movement will be parallel to a line joining A to C. The orientation

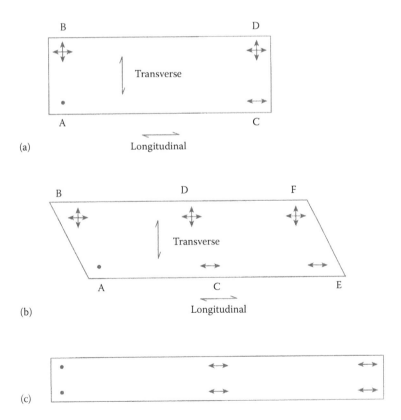

Figure 1.42 Plan views showing articulation of typical bridges: (a) simply supported slab; (b) two-span skewed slab; (c) two-span bridge of small width.

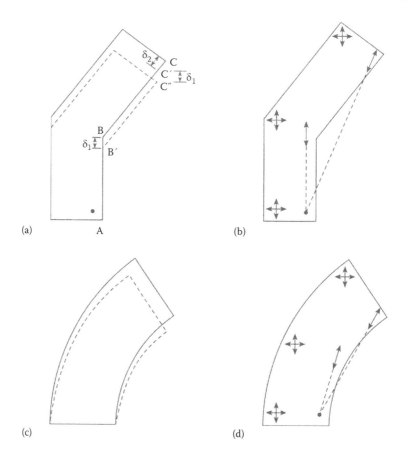

Figure 1.43 Plan views showing articulation of crooked and curved bridges: (a) movement of crooked bridge; (b) articulation to accommodate movement; (c) movement of curved bridge; (d) articulation to accommodate movement.

of bearings that accommodate this movement is illustrated in Figure 1.43b. Similarly, for the curved bridge illustrated in plan in Figure 1.43c, the movements are accommodated by the arrangement of bearings illustrated in Figure 1.43d.

It is expensive to provide bearings capable of resisting an upwards or 'uplift' force. Further, if unanticipated uplift occurs, dust and other contaminants are likely to get into the bearing, considerably shortening its life. Uplift can occur at the acute corners of skewed bridges such as B and E in Figure 1.42b. Uplift can also occur due to applied loading in right (no skew) bridges if the span lengths are significantly different, as illustrated in Figure 1.44. Even with no skew and typical span lengths, differential temperature effects can cause transverse bending, which can result in an uplift as illustrated in Figure 1.45. If this occurs, not only is there a risk of deterioration in the central bearing but also,

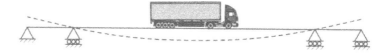

Figure 1.44 Uplift of bearings due to traffic loading in bridge with unequal spans.

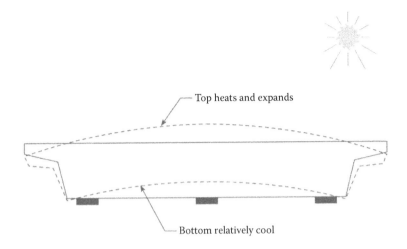

Figure 1.45 Section showing uplift of bearing due to differential temperature effects.

as it is not taking any load, the two outer bearings must be designed to carry the entire load, which renders the central bearing redundant. Such a situation can be prevented by ensuring that the reaction at the central bearing due to permanent loading exceeds the uplift force due to temperature. If this is not possible, it is better to provide two bearings only.

1.6 BEARINGS

There are many types of bearing, and the choice of which type to use depends on the forces and movements to be accommodated and on the maintenance implications. Only a limited number of the more commonly used types are described here. Further details of these and others are given by Lee (1994).

1.6.1 Sliding bearings

Horizontal translational movements can be accommodated using two surfaces that are in contact but are able to slide relative to one another. This is possible using a material with high durability and a very low coefficient of friction such as polytetrafluoroethylene (PTFE). Sliding bearings today generally consist of a stainless-steel plate sliding on a PTFE-coated surface. They can take many forms and are often used in combination with other types of bearing. In some combinations, rotation is facilitated through some other mechanism, and plane sliding surfaces are used, which allow translation only. In other cases, the sliding surfaces are spherical and allow rotation; this form is also referred to as the spherical bearing.

When translation is to be allowed in one direction only, guides are used as illustrated in Figure 1.46. Sliding bearings offer a frictional resistance to movement, which is approximately proportional to the vertical force. Some bearings are lubricated, resulting in a reduced coefficient of friction. However, it is common in such systems for the lubricant to be squeezed out after a number of years, at which time the coefficient returns to the unlubricated value. Whether or not sliding bearings are lubricated, it has been suggested that they be treated as wearing parts that eventually need to be replaced.

Figure 1.46 Guided sliding spherical bearing showing guides at left and right. (Courtesy of Mageba.)

1.6.2 Pot bearings

Pot bearings, illustrated in Figure 1.47, consist of a flat metal cylinder containing an elastomer to which the force is applied by a metal piston. They are frequently used for motorway bridges of moderate span. The elastomer effectively acts as a retained fluid and facilitates some rotation while preventing translation. Thus, pot bearings by themselves are commonly used at the point of translational fixity. They are also used in combination with plane-sliding surfaces to provide free-sliding bearings. By incorporating guides, a pot bearing can be made into a guided sliding bearing.

1.6.3 Elastomeric bearings

When the forces to be resisted are not very high, for example, when bearings are provided under each beam in precast construction, elastomeric bearings (Figure 1.48) can be a very economical alternative to sliding or pot bearings. They are made from rubber and can be in a single layer (for relatively low loading) or in multiple layers separated by metal plates. Elastomeric bearings accommodate rotation by deflecting more on one side than the other (Figure 1.48a) and translation by a shearing deformation (Figure 1.48b). They are considered to be quite durable, except in highly corrosive environments, and require little maintenance.

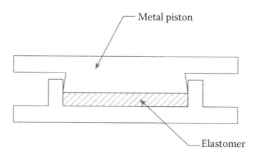

Figure 1.47 Pot bearing.

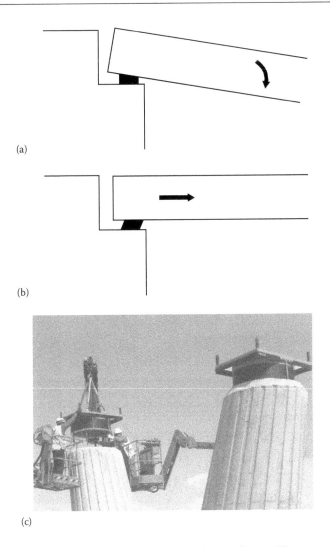

(a)

(b)

(c)

Figure 1.48 Elastomeric bearing: (a) rotation; (b) translation; (c) installation. (Courtesy of Mageba.)

1.7 JOINTS

While bearings in bridges can frequently be eliminated, movements will always occur, which makes it more difficult to avoid joints. Even in integral construction, the movement must be accommodated at the ends of the run-on slabs. However, the number of movement joints being used in bridge construction is decreasing, with the philosophy that all of the associated maintenance implications should be concentrated into as few joints as possible. Joints can be problematic, particularly in road bridges, and frequently leak, allowing salt-contaminated water to wash over the substructures.

Joints should be provided to accommodate all of the movement likely to be encountered in service. The most common types of joints are described below. However, more than one type of joint may be suitable for a particular movement range or site location, and other factors will have to be considered before the final choice of joint type is made.

1.7.1 Buried joint

For movements of 5–20 mm, joints buried beneath road surfacing are possible and, if designed well, can result in a minimum maintenance solution. A typical arrangement is illustrated in Figure 1.49. The material used to span the joint is important; for larger gaps, it is difficult to find a suitable material, which carries the impact loading due to traffic across the gap while facilitating the necessary movement.

1.7.2 Asphaltic plug joint

The asphaltic plug joint is similar to the buried joint in that the gap is protected by road surfacing. However, in this case, the road surfacing over the joint consists of specially formulated flexible bitumen, as illustrated in Figure 1.50. This form has been successfully used for movements of up to 40 mm and is inexpensive to install and replace.

1.7.3 Nosing joint

Very popular in the 1960s and 1970s, the nosing joint, illustrated in Figure 1.51, is no longer favoured in many countries. It can accommodate movements of similar magnitude to the

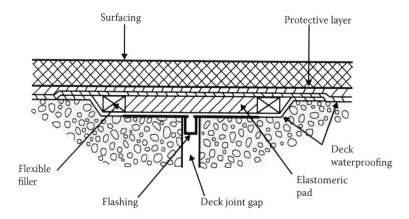

Figure 1.49 Buried joint. (Image provided by the UK Highways Agency, under the Open Government Licence.)

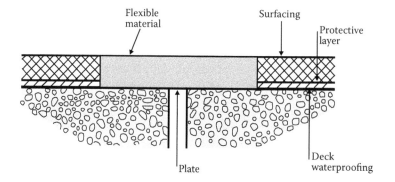

Figure 1.50 Asphaltic plug joint. (Image provided by the UK Highways Agency, under the Open Government Licence.)

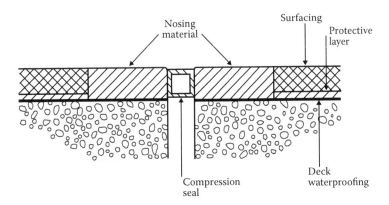

Figure 1.51 Nosing joint. (Image provided by the UK Highways Agency, under the Open Government Licence.)

asphaltic plug joint but has a reputation for frequent failure and leakage. The nosings today are made up of cementitious or polyurethane binders instead of the epoxy mortars popular in the 1970s, which were often found to deteriorate prematurely.

1.7.4 Reinforced elastomeric joint

This type of joint, illustrated in Figure 1.52, comes in various forms from different manufacturers and is supplied in a range of sizes. It has been used for many years with a good success rate. The larger-size elastomeric-type joints tend to create more noise than normal under traffic, but this is usually only a problem if the installation is close to residential property. Resistance to water penetration can be improved by ensuring that the joint is manufactured and supplied in one continuous length.

1.7.5 Elastomeric in metal runners joint

This type of joint, illustrated in Figure 1.53, also comes in various forms from different manufacturers, either as single or multi-element in a range of sizes. Generally, the joints

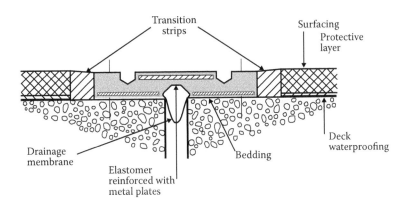

Figure 1.52 Reinforced elastomeric joint. (Image provided by the UK Highways Agency, under the Open Government Licence.)

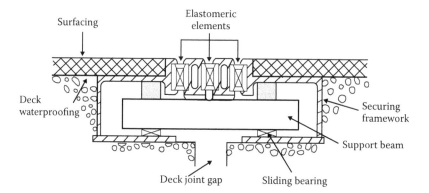

Figure 1.53 Elastomeric in metal runners joint. (Image provided by the UK Highways Agency, under the Open Government Licence.)

are cast in using formed recesses in the deck concrete. Depending on the type of joint used, fixings can be cast-in bolts or site drilled bolts, or it can even be bonded to the deck concrete.

1.7.6 Cantilever comb or tooth joint

These joints can either be purpose-made for a particular installation or be proprietary joints. The gaps between the teeth can become very large, especially on skew bridge decks, and the orientation of the teeth may also be significant in certain circumstances. A typical joint is illustrated in Figure 1.54.

1.8 BRIDGE AESTHETICS

The art of bridge aesthetics is a subjective one, with each designer having his/her own opinions. However, there is generally some common ground, particularly on what constitutes an aesthetically displeasing bridge. Certain bridge proportions, in particular, look better than others, and attention to this can substantially improve the appearance of the structure. The aesthetics of the more common shorter-span bridges are considered in this section. Further details on these and longer-span bridge aesthetics can be found in the excellent book on the subject by Leonhardt (1983).

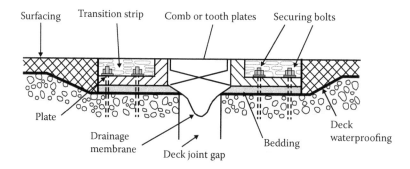

Figure 1.54 Cantilever comb or tooth joint. (Image provided by the UK Highways Agency, under the Open Government Licence.)

Some aspects of aesthetics are common to most bridges. It is generally agreed that the upstand and parapet (refer to Figure 1.1 for terms) are important and that they should be carried through from the bridge to corresponding upstands and parapets in the abutment wing walls as illustrated in Figure 1.55.

This serves to give a sense of continuity between the bridge and its setting, as the eye can follow the line of the bridge from one end to the other. The sun tends to shine directly on upstands, whereas the underside of the main deck tends to remain in shadow (Figure 1.56). This effect can be useful, particularly if the designer wishes to draw attention away from a deep main deck. The effect can be emphasised by casting the upstand in a whiter concrete or by casting the outer surface at an angle to the vertical as illustrated in Figure 1.57. The depth of the upstand and the main deck relative to the span is a critical issue as will be seen in the following sections.

1.8.1 Single-span beam/slab/frame bridges of constant depth

For very short span bridges or culverts, the shape of the opening has a significant influence on the aesthetics. The abutment wing walls also play an important role as can be seen in the example in Figure 1.58. In this example, the shape of the opening is square (span equals height), and the abutment wing walls are large triangular blocks. For such a bridge, the

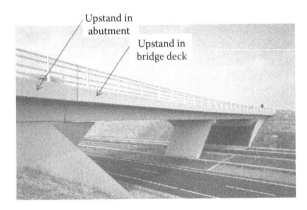

Figure 1.55 Continuity of upstand and parapet. (Courtesy of Roughan and O'Donovan Consulting Engineers.)

Figure 1.56 Shading of main deck relative to upstand. (Courtesy of Roughan and O'Donovan Consulting Engineers.)

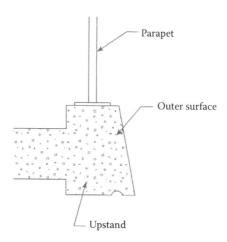

Figure 1.57 Section through upstand.

main deck can be constructed of the same material (e.g. concrete) as the abutment walls. However, it may be difficult to get a good finish with in situ concrete, and if aesthetics are important, it may be better to clad the wing walls in masonry, as illustrated in Figure 1.58a, while leaving the main deck and upstand in concrete.

For a square opening, a relatively deep main deck is recommended such as one fifth of the span. However, this clearly is a matter of opinion and depends on the relative depths of the main deck and the upstand. Three alternatives are illustrated in Figure 1.58. A typical solution is illustrated in Figure 1.58b with a span/upstand depth ratio of 20 and a span/main deck depth ratio of 5. Ratios of 20 and 10 are illustrated in Figure 1.58c for upstand and main deck, respectively, whereas ratios of 10 and 5 are illustrated in Figure 1.58d and a. For a 2 × 1 rectangular opening with wing walls of similar size, a much more slender deck is desirable; span/upstand depth ratios of 20 and a span/main deck depth ratio of 10 are often recommended. For rectangular openings with less pronounced wing walls, a more slender deck is favoured. Typical ratios are illustrated in Figure 1.59a with a span/upstand depth ratio of 40 and a span/main deck depth ratio of 20. The heavier-looking alternative illustrated in Figure 1.59b has ratios of 60 and 10. It can be seen in the latter that the upstand appears too thin and/or the deck too deep.

Leonhardt points out that scale is important as well as proportion. This is illustrated in Figure 1.60a, where people and traffic are close to the structure, which is large relative to their size. (In this structure, a parapet wall is integral with the upstand, making it look deeper than necessary.) A structure with similar proportions looks much better in Figure 1.60b as it is smaller and is more likely to be viewed from a distance.

1.8.2 Multiple spans

The relative span lengths in multi-span bridges have a significant effect on the appearance. For aesthetic reasons, it is common practice in a three-span construction to have the centre span greater than the side spans, typically by 25%–35% as illustrated in Figure 1.61. This can be convenient as the principal obstruction to be spanned is often in the central part of the bridge. When the ground level is lower at the centre, as illustrated in Figure 1.61, this proportioning also tends to bring the relative dimensions of the rectangular openings closer, which has a good aesthetic effect. The bridge illustrated is typical with a main span/upstand depth ratio of 40 and a span/deck depth ratio of 20. As for single-span bridges, the upstand

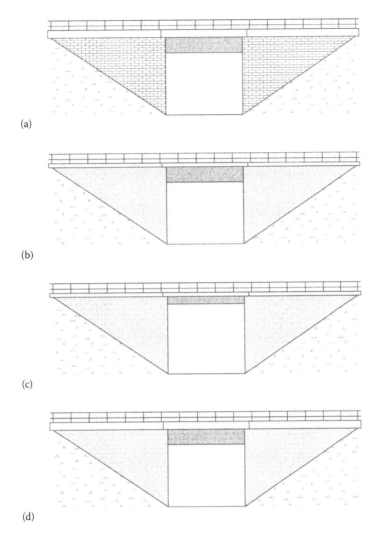

(a)

(b)

(c)

(d)

Figure 1.58 Square opening with alternative span/upstand and span/main deck depth ratios: (a) 10 and 5 with brick wing walls; (b) 20 and 5; (c) 20 and 10; (d) 10 and 5.

is continuous from end to end, effectively tying the bridge together. An open parapet is also used in the bridge of Figure 1.61 to increase its apparent slenderness.

Varying the depth of bridges allows the depth to be increased at points of maximum moment. This complicates the detailing but makes for an efficient light structure and tends to look very well. When a road or rail alignment is straight, straight haunches are possible as illustrated in Figure 1.62a: the depth is increased at the points of maximum (hogging) moment. Straight haunches are considerably cheaper than curved ones, both in terms of shuttering and reinforcement details. However, they are not as aesthetically pleasing as a curved profile, illustrated in Figure 1.62b and c. When alignments are curved, curved decks are strongly favoured over straight ones. In Figure 1.62b, the bottom of the bridge has a different radius of curvature from the top. This has the effect of increasing the depth where it is needed in the longer spans. In Figure 1.62c, curved haunches are provided over the internal supports to give extra depth at the points of maximum hogging moment.

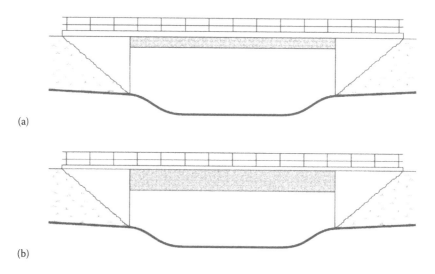

(a)

(b)

Figure 1.59 Rectangular opening with small wing walls: (a) slender deck and deep upstand; (b) deep deck and slender upstand.

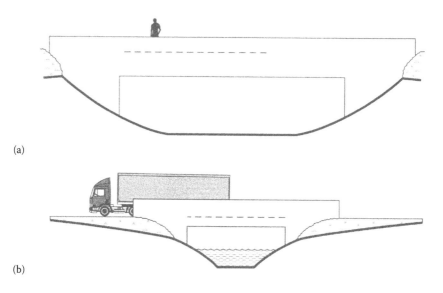

(a)

(b)

Figure 1.60 The influence of scale on appearance: (a) large structure near the viewer looks heavy; (b) small structure remote from the viewer looks better.

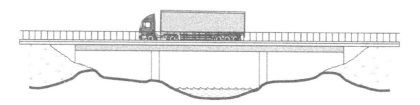

Figure 1.61 Three-span bridge with good proportions.

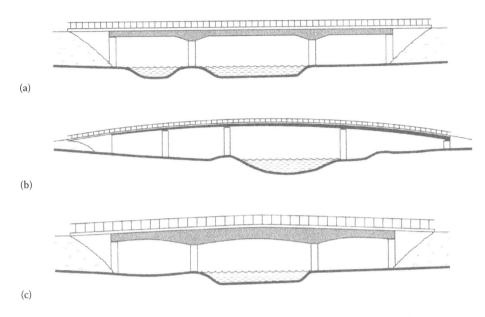

(a)

(b)

(c)

Figure 1.62 Variable depth bridges: (a) straight haunches; (b) curved alignment achieved using two curves of differing radius; (c) curved haunches.

Chapter 2

Bridge loading

2.1 INTRODUCTION

For bridges, it is often necessary to consider phenomena that would normally be ignored in building structures. For example, effects such as differential settlement of supports frequently need to be considered by bridge designers while generally being ignored by designers of buildings. These and other more common forms of bridge loading are considered in this chapter. The various types of loading that need to be considered are summarised in Table 2.1. Some of these are treated in greater detail in the following sections, as indicated in the third column of Table 2.1. Other types of loading that may occur but that are not considered here are exceptional loads (such as snow) and construction loads. Another source of loading is earth pressure on substructures. This is considered in Chapter 4 in the context of integral bridges. Two codes of practice are referred to in this chapter, namely, Eurocode 1 (EN 1991-1 2002 and EN 1991-2 2003) and the American AASHTO standard (AASHTO 2010).

Dead loads consist of permanent gravity forces due to structural elements and other permanent items such as parapets and road surfacing. Imposed traffic loads consist of those forces induced by road or rail vehicles on the bridge. The predominant effect is the vertical gravity loading, including the effect of dynamics. However, horizontal loading due to braking/traction and centrifugal effects in curved bridges must also be considered. Where footpaths or cycle tracks have been provided, the gravity loading due to pedestrians/cyclists can be significant.

Thermal changes can have significant effects, particularly in frame and arch bridges. Two types of thermal effect are considered. Uniform changes in temperature cause a bridge to expand or contract. If these changes in length are constrained, stresses are generated. Differential temperature changes happen when parts of the bridge cross section heat up or cool down more than the other parts. This causes curvature and will generate bending moment if constrained.

Differential (i.e. relative) settlement of supports can induce significant bending in continuous beam or slab bridges, as will be demonstrated in Chapter 3. The conventional approach to this problem is to estimate the maximum differential displacement of supports and to impose this in an analysis of the bridge. The Eurocode on Geotechnical Design, EN 1997 (2004), points out that predictions of this type where the displacement is estimated, ignoring soil/structure interaction, tend to be over-predictions. The AASHTO code specifies that differential settlement should be considered and gives guidance for the case of piled foundations.

The horizontal loading due to impact from collisions with errant vehicles can be quite significant for some bridge elements. For all bridges subject to traffic loading, there is an additional vertical component of loading due to vehicle/bridge dynamic interaction.

Table 2.1 Summary of bridge loads and effects

Load type	Description	Section
1. Dead	Gravity loading due to weights of structural and non-structural parts of bridge	2.2
2. Imposed traffic	Loading due to road and rail vehicles	2.3
3. Pedestrian and cycle track	Gravity loading due to non-vehicular traffic	2.3
4. Shrinkage and creep	Shrinkage and creep effects in concrete structures	2.4
5. Thermal	Uniform and differential changes in temperature	2.5
6. Differential settlement	Relative settlement of supporting foundations	–
7. Impact	Impact loading due to collision with errant vehicles	2.6
8. Dynamic effects	Effect of bridge vibration	2.7
9. Wind	Horizontal loading due to wind on parapets, vehicles and the bridge itself	–
10. Prestress	Effect of prestress on indeterminate bridges	2.8

Vibration is generally more significant in slender light bridges. In practice, this may be a pedestrian bridge or a long-span road or rail bridge, where the natural frequency of the bridge is at a level that can be excited by traffic or wind. In pedestrian bridges, it should be ensured that the natural frequency of the bridge is not close to that of walking or jogging pedestrians.

In addition to its ability to induce vibration, wind can induce static horizontal forces on bridges. The critical load case generally occurs when a convoy of high vehicles is present on the bridge, resulting in a large vertically projected area. Wind tends not to be critical for typical road bridges that are relatively wide but can be significant in elevated railway viaducts when the vertically projected surface area is large relative to the bridge width.

Prestress is not a load as such but a means by which applied loads are resisted. However, in indeterminate bridges, it is necessary to analyse to determine the effect of prestress; thus, it is often convenient to treat it as a form of loading. The methods used are very similar to those used to determine the effects of temperature and shrinkage changes.

The AASHTO code specifies a range of combinations of load that should be checked. For example, the 'Strength I' check combines traffic load factored by 1.75 with factored dead load and thermal effects but without any wind load. The 'Strength V' check considers the same effects plus wind but factors the traffic load by just 1.35 to allow for the reduced probability of both phenomena being at extreme levels at the same time. The Eurocode uses combination factors, ψ_o, to allow for this reduced probability.

2.2 DEAD LOADING

For bridge structures, dead loading is the gravity loading due to the structure and other items permanently attached to it. In the old British Standard, BS5400, there was a subdivision of this into dead loading and superimposed dead loading. Gravity loading from non-structural parts of the bridge was considered to be superimposed. Such items are long term but might be changed during the lifetime of the structure. An example of superimposed dead load is the weight of ballast in railway bridges. There is clearly always going to be ballast, thus, it is a permanent source of loading. However, it is possible that the depth will increase at certain times in the life of the bridge. Because of such uncertainty, superimposed dead load was assigned a higher factor of safety than dead load.

In the Eurocode, all permanent gravity loading is considered to be dead load, and this uncertainty in some non-structural items is treated by adding a percentage to the nominal value. For example, the Eurocode specifies a variation of up to 30% in the nominal depth of ballast (the United Kingdom National Annex* to the Eurocode includes some small differences to this clause). Apart from these special allowances for non-structural items, dead load is simply calculated as the product of volume and material density. For prestressed concrete bridges, it is important to remember that an overestimate of the dead load can result in an underestimate of the transfer stresses. Thus, dead load should be estimated as accurately as possible rather than simply rounded up.

Bridges are unusual among structures in that a high proportion of the total loading is attributable to dead load. This is particularly true of long-span bridges. In such cases, steel is common due to its high strength-to-weight ratio. For shorter spans, concrete or composite steel beams with concrete slabs are the usual materials. In some cases, lightweight concrete has been successfully used to reduce the dead load.

2.3 IMPOSED TRAFFIC LOADING

Bridge traffic can be vehicular, rail or pedestrian/cycle or indeed any combination of these as described in the following subsections.

2.3.1 Pedestrian traffic

Road and rail traffic are considered in subsections below. While pedestrian/cycle traffic loading on bridges is not difficult to calculate, its importance should not be underestimated. Bridge codes commonly specify a basic intensity for pedestrian crowd loading (e.g. 5 kN/m² in the Eurocode), which may be reduced for long bridges. When a structural element supports both pedestrian and traffic loading, a lesser intensity is allowed to reflect the reduced probability of both traffic and pedestrian loading reaching extreme values simultaneously. In Europe, this can be specified in the National Annex – the Eurocode recommends 3 kN/m². The AASHTO code specifies a loading of 3.6 kN/m² (0.075 kip/ft²) for elements supporting both pedestrians and traffic loading.

2.3.2 Nature of road traffic loading

Road bridges are designed to take the weights of truck traffic. Short-span bridges are governed by small numbers of extremely heavy trucks, with an allowance for dynamic vehicle/ bridge interaction. Long-span bridges, on the other hand, are governed by slow-moving congested traffic conditions.

While some truck-weighing campaigns have been carried out in the past, there was a scarcity of good unbiased data on truck weights until recent years. In the past, sampling was carried out by taking trucks from the traffic stream and weighing them statically on weighbridges. There are two problems with this as a means of collecting statistics on truck weights. In the first place, the quantity of data collected is relatively small, but more importantly, there tends to be a bias, as drivers of illegally overloaded trucks learn quickly that weighing is taking place and take steps to avoid that location. In recent years, the situation has improved considerably with improvements in weigh-in-motion (WIM) technology, which allows all trucks passing a sensor to be weighed while they travel at full highway

---

* The Eurocodes allow some variations between countries; these are defined in each country's National Annex.

speed. WIM technology has resulted in a great increase in the availability of truck weight statistics, and there is now a much better understanding of the nature of extreme traffic loading.

Measured WIM data can be used to calculate 'load effects' (LEs), that is, bending moments or shear forces for any particular bridge. With years of WIM records, data are made more manageable by considering only the maximum LE in a given block of time such as a day or a week. A cumulative distribution function for maximum-per-day data – Figure 2.1a – plots the LE against its probability of not being exceeded, F_{NE}. Hence, the extremely rare LEs of interest correspond to non-exceedance probabilities approaching unity. The cumulative probabilities in this graph are generally rescaled to better illustrate the extremely rare data. These rescaled cumulative distribution graphs are known as probability paper plots. Figure 2.1b is a probability paper plot made by plotting the same data as Figure 2.1a on a double-log (Gumbel) scale – this is known to be suitable for illustrating extreme value data of this type. The non-exceedance probability values corresponding to 75- and 1000-year return periods are shown in this graph for reference (AASHTO and the Eurocode use 75- and 1000-year return periods, respectively).

Allowing for public holidays, there are approximately 250 working days in a year. Hence, a return period of, say, 75 years corresponds to an event that occurs just once in 18,750 days. As the data are maximum-per-day, this corresponds to a probability of non-exceedance of $1 - 1/18,750 = 0.99995$. It can be seen in Figure 2.1b that, for this LE and site, the 75-year LE is approximately 10,000 kNm.

It is generally accepted that there are two kinds of vehicles: standard vehicles, which are within the normal legal weight limits, and special-permit vehicles, which have permission to travel with weights outside these limits. There are many statistical studies in the literature (Nowak and Szerszen 1998; Cremona 2001; O'Connor and OBrien 2005; O'Connor and Eichinger 2007; Caprani and OBrien 2010; OBrien and Enright 2011; Enright and OBrien 2013) where WIM data are used to calculate the characteristic bending moment or shear force for a given return period. Unfortunately, permit-issuing databases are not generally

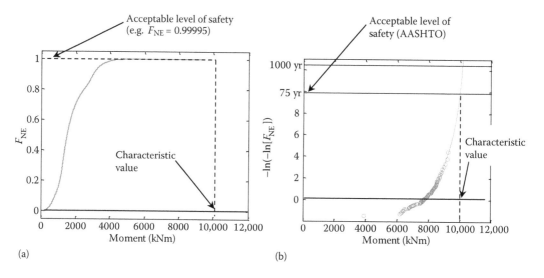

Figure 2.1 Maximum-per-day LE data for mid-span bending moment in a 40-m-long bridge using WIM data from Arnhem, the Netherlands (F_{NE} = probability of non-exceedance): (a) cumulative distribution function; (b) probability paper plot.

linked to WIM databases, so it is difficult to separate standard trucks from those with a special permit. There are also many incidences of illegal overloading; if a vehicle is outside the legal weight limits, it may or may not have a permit. As a result, many statistical studies consider all vehicles, standard and special permit, together.

Enright et al. (2014) have proposed a set of WIM data filtering rules to separate vehicles that are unlikely to have a permit – apparent standard vehicles – from those that are likely to have one – apparent permit vehicles. If data are filtered by vehicle weight, it excludes the possibility of vehicles being illegally overloaded; thus, they applied filters based on numbers of axles and axle spacings only. The results, using the same data as Figure 2.1, are illustrated in Figure 2.2. It can be seen that, for a given return period, the apparent standard vehicles are much less critical than the apparent permit vehicles. For a 75-year return period, the characteristic maximum LE caused by an apparent standard vehicle is approximately 7400 kNm. The corresponding figure for the apparent permit vehicle data is approximately 10,000 kNm. It could be argued that permit vehicles are controlled, unlike standard vehicles, and that it is not necessary to extrapolate in this way to find a characteristic maximum LE. If this is accepted, and the bridge has a capacity to resist this LE of, say, 8000 kNm, then it would be deemed safe for standard vehicles, and permit vehicles would be prohibited that result in an LE greater than 8000 kNm. This approach does, of course, assume that all apparent permit vehicles can be controlled.

Using WIM data unsorted by permits, critical loading scenarios for short- to medium-span bridges (OBrien and Enright 2011, 2013; Enright and OBrien 2013) tend to consist of either (a) one extremely heavy vehicle or (b) a very heavy vehicle in one lane with a standard vehicle in the adjacent lane. The very heavy vehicles are either large mobile cranes or 'low loaders' that would be expected to have a permit (Figure 2.3). In Europe, these mobile cranes are typically between 100 and 110 tonnes, with closely spaced and heavily loaded axles. In North America, mobile cranes are usually carried on a larger number of less heavily loaded axles. Low loaders have a gap of approximately 11 m near the middle and large numbers of axles on either side. Unlike mobile cranes, there is no known upper limit to their weight at this time – as ever larger databases of vehicles are searched, ever heavier low loaders are being found.

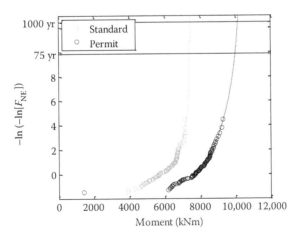

Figure 2.2 Probability paper plot of maximum-per-day LE data of Figure 2.1, sorted by whether the vehicle is likely to have a permit or not (F_{NE} = probability of non-exceedance).

(a)   (b)

Figure 2.3 Critical vehicles for bridge loading: (a) mobile crane; (b) low loader. (Courtesy of Rijkswaterstaat, Ministry of Infrastructure and the Environment, the Netherlands.)

2.3.3 Code models for road traffic

The Eurocode considers 'Normal' and 'Abnormal' loading. The Normal Load model (LM1) represents the worst combination of standard vehicles expected to occur in 5% of bridges, just once in a 50-year lifetime. This approximates to the loading scenario with a (50/0.05 =) 1000-year return period. The Eurocode gives suggestions for abnormal or special permit vehicles in an annex, but national authorities are permitted to specify what they wish. Combinations of normal traffic and an abnormal vehicle must be considered in bridge design. The abnormal load in the Eurocode is generally taken to replace the normal loading throughout the length of the vehicle and for a distance of 25 m before and after it. Normal load is placed throughout the remainder of the lane and in the other lanes.

A similar situation exists in the United States. The HL93 load model in the AASHTO code is deemed to represent the worst combination of loading from standard and routine permit vehicles in a 75-year return period. While this may sound like a much lower return period than that used in Europe, return periods are commonly measured on a double log scale; thus, the difference between 75 and 1000 is not as much as it seems. Abnormal loading in the United States is dealt with at the state level – some states require their bridges to be designed to carry a notional load model in addition to HL93.

Bridge traffic loading is applied to notional lanes that are independent of the actual lanes delineated on the road. In the Eurocode, the road width is divided into a number of notional lanes, each 3 m wide. The outstanding road width between kerbs, after removing these lanes, is known as the 'remaining area'. The AASHTO code also specifies notional lanes, generally with a fixed width of 3.66 m (12 ft.), with the number of lanes rounded down to the nearest integer number.

The Eurocode normal load model consists of uniform loading and a tandem of two axles (four wheels) in each lane as illustrated in Figure 2.4a. In addition, there is uniform loading in the remaining area. While there are a number of factors that can vary between road classes and between countries, the standard combination is a load intensity of 9 kN/m^2 in lane no. 1 and 2.5 kN/m^2 elsewhere. (Notably, in the UK National Annex, these loadings are factored to give 5.5 kN/m^2 in all lanes.) The four wheels of the tandems together weigh 600, 400 and 200 kN for lanes 1, 2 and 3, respectively. All of these loads are deemed to include an allowance for dynamics.

The AASHTO HL93 notional load model, illustrated in Figure 2.4b, consists of a truck or tandem axle (whichever is worse) plus a uniformly distributed lane loading. Except for fatigue and for the particular case of joints, a factor of 1.33 is applied to the truck or tandem part of the model to allow for the dynamic interaction between the vehicle and the bridge.

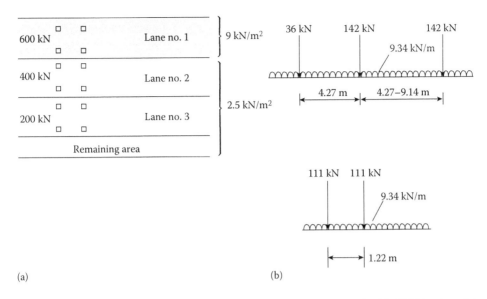

Figure 2.4 'Normal' road traffic loading: (a) Eurocode normal loading; (b) AASHTO HL93 load model.

As illustrated in Figure 2.4a, the loadings in the Eurocode are much heavier in one lane than in the others. This is to account for the reduced probability of extreme loading in many lanes simultaneously. The AASHTO code also allows reductions in lane loading for multi-lane bridges, in this case, with 'multiple presence' factors. The multiple presence factors are 1.2, 1.0, 0.85 and 0.65 for one, two, three and more than three lanes, respectively. Thus, for example, if two notional lanes are considered, full HL93 should be applied in each, whereas if three lanes are considered, this can be reduced to 0.85 × HL93 in each. For continuous beam/slab bridges, two trucks, with some further reduction factors, can be present simultaneously on the same lane of the bridge.

The Eurocode does not consider spans beyond 200 m, and while the AASHTO code does not specify an upper limit, it is stated in the commentary that it has been calibrated for bridges up to 61 m and that there have been 'spot checks on a few bridges' for lengths of up to 183 m. The standard notional load models tend to be conservative for longer bridges, and specialist statistical studies of traffic loading are advised.

2.3.4 Imposed loading due to rail traffic

The modelling of railway loading is considerably less complex than road traffic loading as the transverse location of the load is fixed – the train can generally be assumed to remain on the tracks. However, there are some aspects of traffic loading that are specific to railway bridges and must be considered. The weights of railway carriages can be much better controlled than those of road vehicles, with the result that different load models are possible depending on the railway line on which the bridge is located. However, bridges throughout a rail network are generally designed for the same normal load model. The standard Eurocode normal load model – Load Model 71 – consists of four vertical point loads at 1.6 m intervals of magnitude 250 kN each and uniform loading of intensity 80 kN/m both before and after them. In addition, the Eurocode provides for an alternative SW/2 load model for heavy-rail traffic.

In the United States, the American Railway Engineering and Maintenance-of-way Association (AREMA) manual for Railway Engineering is generally adopted by states. The AREMA Cooper E load model is specified in that manual, even though the axle weights and spacings no longer match those of today's locomotives. There are a number of alternative gross weights, each following the same configuration of axle spacings. The Cooper E80 is popular with a gross weight of 5053 kN over 33 m, so named because the locomotive drive axle weights are 80 kips (356 kN).

On passenger transit 'light rail' systems, less onerous load models can be applied than those specified in the Eurocode and the AREMA manual. These tend to be network-specific, and notional load models are developed on an ad hoc basis.

The static loads specified for the design of railway bridges must be increased to take account of the dynamic effect of carriages arriving suddenly on the bridge. This factor is a function of the permissible train speed and of the natural frequency of the bridge. Railway tracks on grade are generally laid on ballast. On bridges, tracks can be laid on a concrete 'track slab', or the bridge can be designed to carry ballast and the track laid on this. There are two disadvantages to the use of track slabs. When used, an additional vertical dynamic load is induced by the change from the relatively 'soft' ballast support to the relatively hard track slab. This effect can be minimised by incorporating transition zones at the ends of the bridge with a ballast of reducing depth. The other disadvantage to the use of track slabs depends on the method used to maintain and replace ballast. If this is done using automatic equipment, a considerable delay can be caused by the need to remove the equipment at the start of the bridge and to reinstall it at the other end.

Another aspect of loading specific to railway bridges is the rocking effect. It is assumed for design purposes that more than half of the load (approximately 55%) can be applied to one rail, while the remainder (approximately 45%) is applied to the other. This can generate torsion in the bridge.

Horizontal loading due to braking and traction is more important in railway bridges than in road bridges as the complete train can brake or accelerate at once. While it is possible in road bridges for all vehicles to brake at once, it is statistically much less likely. Longitudinal horizontal loading in bridges can affect the design of bearings and can generate bending moment in substructures and throughout frame bridges.

2.4 SHRINKAGE AND CREEP

Shrinkage is the reduction in the length of a concrete member that occurs as the concrete sets and as moisture is lost over time. Creep is the increase in deformation over time that occurs in the presence of a sustained stress. In the context of bridges, creep is often associated with prestressed concrete and manifests itself as a continuing shortening over time. However, there are also creep effects that result from the changes in geometry that happen during construction. Creep and shrinkage are related phenomena. They are both a function of the composition of the concrete and of the gradual movements of moisture through it that happen over time. They are both influenced by the surface area per unit volume or, in cross section, the surface perimeter per unit area. A high ratio means a greater loss of moisture and a greater tendency to shrink and creep. Creep is also influenced by the maturity of the concrete when the stress is first applied – early application of stress results in greater creep in the long term.

Force causes stress in a member, and excessive applied stress can cause failure. Shrinkage and creep are applied strains, that is, applied deformations, which is different to applied stresses. If a member is free to deform, then an applied strain only causes deformation. For example, if a concrete member is free at both ends and it shrinks, it becomes shorter than before, but no stress is induced, and it is no nearer to failure. On the other hand, if a member is restrained against movement, then an applied strain does generate stress, and it can fail. An example of this is a concrete member fixed against movement at each end. Shrinkage in this situation can cause cracking and failure in tension.

2.4.1 Shrinkage

The Eurocodes separate total shrinkage strain into two components: autogenous shrinkage strain and drying shrinkage strain. Autogenous shrinkage strain occurs as concrete continues to set over time. It is therefore short term, and most of it occurs in the first few weeks after the concrete is cast. Drying shrinkage strain results from the migration of moisture through the hardened concrete over time. This is much more long term and takes years to be substantially complete.

For autogenous shrinkage strain, the Eurocodes specify a simple linear function of concrete strength – stronger concrete has greater autogenous shrinkage. Drying shrinkage strain is given as a function of concrete strength, relative humidity at the site and the surface perimeter per unit area. Drying shrinkage is less when concrete is strong, humidity is high and surface perimeter per unit area is low.

The AASHTO code does not use the terms autogenous and drying shrinkage but does point out that strains are relatively rapid in the first several weeks and much slower thereafter. As in the Eurocodes, total shrinkage strain is given as a function of concrete strength, relative humidity and surface perimeter per unit area.

2.4.2 Creep

Creep strain is given by $\varphi\sigma_{el}/E$, where φ is the creep coefficient, σ_{el} is the linear elastic stress and E is the modulus of elasticity. The Eurocodes give the creep coefficient as a function of the concrete strength and its tangent modulus, which, in turn, is a function of the cement type and the age of concrete at the time of loading. In broad terms, creep can be reduced by delaying the first loading or using stronger concrete. The AASHTO code uses a simpler formula that gives the creep coefficient as a function of concrete strength and the age of concrete at the time of loading. Unlike the Eurocodes, it also has factors that account for relative humidity and surface perimeter per unit area.

2.5 THERMAL LOADING

There are two thermal effects that can induce stresses in bridges. The first is a uniform temperature change, which results in an axial expansion or contraction, and the second is a differential temperature, which causes curvature. Like shrinkage, temperature change is an applied strain, not stress. If unrestrained, a change in temperature causes movement, that is, strain and not stress. It is only when the movement is restrained, such as in a frame or arch bridge, that temperature change can generate distributions of stresses in the form of axial force, bending moment and shear force.

Differential changes in temperature occur when, for example, the top of a beam heats up relative to the bottom. This causes the top to expand relative to the bottom. If unrestrained, as is the case for a simply supported beam, this causes curvature, and the centre of the beam rises up relative to the ends. If the curvature is restrained, as happens in two-span continuous beams, differential temperature change causes bending moment and shear force.

2.5.1 Uniform changes in temperature

Uniform changes in temperature result from sustained periods of hot or cold weather in which the entire depth of the deck undergoes an increase or decrease in temperature. The national annexes to the Eurocode give contour plots of maximum and minimum air shade temperature, T_{max} and T_{min} – see Figure 2.5 for example. The difference between air shade temperature and the effective temperature within a bridge, $T_{e,max}$ or $T_{e,min}$, depends on the thickness of surfacing and on the form of construction. For thermal purposes, there are three forms of construction:

Type 1 – Steel: orthotropic steel deck supported by box girder, truss or plate girders
Type 2 – Composite: concrete deck supported by steel-box girder or plate girders
Type 3 – Concrete: solid concrete slab, beam-and-slab or box girder construction

Typical thicknesses for the surfacing are 40 mm for type 1 and 100 mm for types 2 or 3. For these values, the effective uniform bridge temperatures are as given in Table 2.2. For other surface depths, adjustment formulas are specified in the national annexes. It can be seen that in very warm weather, the effective temperature in a steel bridge can be considerably higher than the ambient temperature, and that it can be colder in very cold weather. Concrete, on the other hand, tends to store heat, and the effective temperatures are always a few degrees higher than the air shade temperatures.

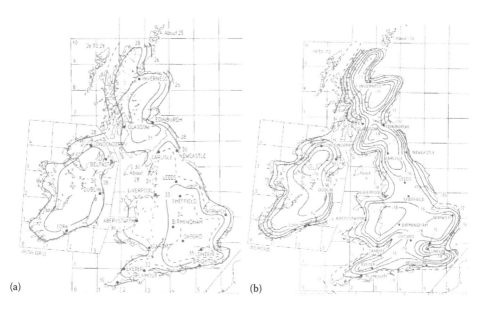

Figure 2.5 Isotherms of air shade temperature for the United Kingdom and Ireland: (a) maximum, T_{max}; (b) minimum, T_{min}. (From UK National Annex to Eurocode 1: Actions on structures, Part 1-5: General actions – Thermal actions, Figures NA.1 and NA.2, NA to BS EN 1991-1-5:2003.)

Table 2.2 Effective uniform bridge temperatures ($T_{e,max}$ and $T_{e,min}$) in °C as a function of air shade temperatures (T_{max} and T_{min})

Type 1 (Steel)	Type 2 (Composite)	Type 3 (Concrete)
$T_{e,max} = T_{max} + 16$	$T_{e,max} = T_{max} + 4$	$T_{e,max} = T_{max} + 2$
$T_{e,min} = T_{min} - 3$	$T_{e,min} = T_{min} + 4$	$T_{e,min} = T_{min} + 7$

The AASHTO code specifies two procedures: A and B. Procedure A is the older, established approach and is simpler to apply. Procedure B is more sophisticated and is only allowed for composite and concrete bridges.

For Procedure A, the United States is divided into regions with 'moderate' and 'cold' climates, where cold is defined as a location where the average temperature through the day is below freezing for 14 or more days of the year. In 'cold' climates, metal bridges must be designed for temperatures in the range −34°C to 49°C and concrete bridges for temperatures in the range −18°C to 27°C. For 'moderate' climates, the same upper values are used, but the lower values are less extreme at −18°C and −12°C for metal and concrete bridges, respectively. The AASHTO Procedure B is similar to the Eurocode approach, but the effective uniform temperatures, denoted here as $T_{MaxDesign}$ and $T_{MinDesign}$, are given directly in contour plots for each bridge type.

It is important in bridge construction to establish a baseline for the calculation of uniform temperature effects, that is, the temperature of the bridge at the time of construction. It is possible to control this baseline by specifying the permissible range of temperature in the structure at the time of completion of the structural form. Completion of the structural form could be the process of setting the bearings or the making of a frame bridge integral. In concrete bridges, high early temperatures can result from the hydration of cement, particularly for concrete with high cement content. Resulting stresses in the period after construction will tend to be relieved by creep, although little reliable guidance is available on how this might be allowed for in design. Unlike in situ concrete bridges, those made from precast concrete or steel will have temperatures closer to ambient during construction.

As is discussed in Chapter 4, integral bridges undergo repeated expansions and contractions due to daily or seasonal temperature fluctuations. After some time, this causes the backfill behind the abutments to compact to an equilibrium density. In such cases, the baseline temperature is clearly a mean temperature that relates to the density of the adjacent soil.

EXAMPLE 2.1   UNIFORM TEMPERATURE

The concrete slab bridge of Figure 2.6 will be constructed in Birmingham in the United Kingdom. It has been specified that the bearings will be fixed in place when the air shade temperature is in the range of 8°C to 25°C. Calculate the maximum movement at the bearings, given that the coefficient of thermal expansion is 12 × 10⁻⁶/°C.

Figure 2.5 gives the extremes of air shade temperature that could occur in the lifetime of the bridge. Rounding up gives a maximum of 34°C, and rounding down gives a minimum of −18°C. For a type 3 bridge, the effective temperatures are, from Table 2.2,

$T_{e,max} = T_{max} + 2 = 36°C$
$T_{e,min} = T_{min} + 7 = -11°C$

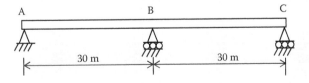

Figure 2.6 Two-span bridge of Example 2.1.

Hence, the greatest possible rise in temperature is

$$36 - 8 = 28°C$$

and the greatest fall is

$$25 - (-11) = 36°C$$

The coefficient of thermal expansion is a material property that gives the change in strain per degree change in temperature. This bridge is free to expand to the right from the fixed support at A. The maximum expansion will be at C, 60 m from that fixed point. The maximum expansion at C is therefore

$$(12 \times 10^{-6}/°C)(28°C)(60 \text{ m}) = 0.020 \text{ m}$$

and the greatest contraction is

$$(12 \times 10^{-6}/°C)(36°C)(60 \text{ m}) = 0.026 \text{ m}$$

Hence, the range of movement is 46 mm, and the bearing should be set to allow more contraction than expansion. (Note that temperature changes are not the only source of movement – concrete bridges may also have movement due to shrinkage and creep.)

There is sometimes pressure on designers to limit thermal movement in order to facilitate the use of a more durable joint such as a buried or asphaltic plug joint (see Section 1.7). The range of movement can sometimes be reduced by changing the articulation. In Example 2.1, fixing the bearing at B and allowing equal movements at A and C would halve the maximum and minimum movements. Another way to reduce the movements is to reduce the coefficient of thermal expansion. The coefficient of thermal expansion of concrete is strongly influenced by the aggregate from which it is made. By specifying a different kind of aggregate, it is possible to reduce the coefficient and hence the range of movement.

2.5.2 Differential changes in temperature

In addition to uniform changes in temperature, bridges are subjected to differential temperature changes on a daily basis, such as in the morning when the sun shines on the top of the bridge, heating it up faster than the interior. The reverse effect tends to take place in the evening, when the deck is warm in the middle but is cooling down at the top and bottom surfaces. The Eurocode allows two approaches to differential temperature: Approach 1 involves a linear variation in temperature through the depth, whereas Approach 2 specifies a more complex non-linear distribution (such as that illustrated in Figure 2.9). The national annex in the United Kingdom specifies that only Approach 2 shall be used. The AASHTO code is similar to Eurocode's Approach 2; it specifies a non-linear distribution of temperature whose parameters vary by bridge type and by region in the United States.

Two distributions of differential temperature are specified in both codes, one corresponding to the heating-up period and one corresponding to the cooling-down period. These distributions can be resolved into axial, bending and residual effects as will be illustrated in the following examples. Transverse temperature differences can occur when one face of a superstructure is subjected to direct sun while the opposite side is in the shade. This effect can be particularly significant when the depth of the superstructure is great. Cracking of reinforced concrete members reduces the effective cross-sectional area and second moment of area. If cracking is ignored, the magnitude of the resulting thermal stresses can be significantly overestimated.

The effects of both uniform and differential temperature changes can be determined using the method of 'equivalent loads'. A distribution of stress is calculated corresponding to the specified change in temperature. This is resolved into axial, bending and residual distributions as will be illustrated in the following examples. The corresponding forces and moments are then readily calculated. Methods of analysing to determine the effects of the equivalent loads are described in Chapter 3.

EXAMPLE 2.2   DIFFERENTIAL TEMPERATURE CHANGE IN SOLID SECTION

The bridge beam illustrated in Figure 2.7 is subjected to the differential increases in temperature shown. It is required to determine the effects of the temperature change if it is simply supported on one fixed and one sliding bearing. The coefficient of thermal expansion is 12×10^{-6}, and the modulus of elasticity is 35,000 N/mm².

The applied temperature distribution is converted into the equivalent stress distribution of Figure 2.8a by multiplying by the coefficient of thermal expansion and the modulus of elasticity. There is an 'equivalent' axial force and bending moment associated with any distribution of temperature. The equivalent axial force can readily be calculated as the sum of products of stress and area:

$$F = \tfrac{1}{2}(5.04)(600 \times 300) + \tfrac{1}{2}(2.10)(600 \times 200)$$
$$= 579,600 \text{ N}$$

This corresponds to a uniform axial stress of $579,600/(600 \times 1200) = 0.81$ N/mm² as illustrated in Figure 2.8b. However, this beam is supported on a sliding bearing at one end and is therefore free to expand. Thus, there is in fact no axial stress but a strain of magnitude $0.81/35,000 = 23 \times 10^{-6}$.

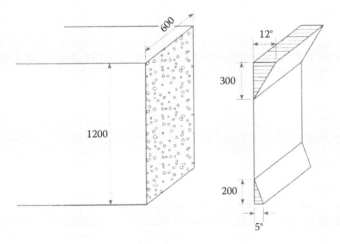

Figure 2.7 Beam subject to differential temperature change.

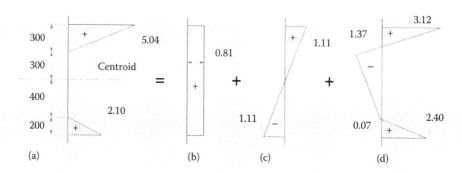

Figure 2.8 Components of imposed stress distribution: (a) total distribution; (b) axial component; (c) bending component; (d) residual stress distribution.

The equivalent bending moment is found by taking moments about the centroid of each component of force:

$$M = -\tfrac{1}{2}(5.04)(600 \times 300)(500) + \tfrac{1}{2}(2.10)(600 \times 200)(533)$$
$$= -160 \times 10^6 \text{ N mm}$$

The corresponding extreme fibre stresses are

$$\frac{My}{I} = \pm \frac{(160 \times 10^6)(600)}{(600 \times 1200^3/12)} = \pm 1.11 \text{ N/mm}^2$$

as illustrated in Figure 2.8c. As the beam is simply supported, it is free to bend and there is in fact no such stress. Instead, a strain distribution is generated, which varies linearly in the range $\pm 1.11/35{,}000 = \pm 32 \times 10^{-6}$. The difference between the applied stress distribution and that which results in axial and bending strains is trapped in the section and is known as the residual stress distribution, illustrated in Figure 2.8d. It is found simply by subtracting Figure 2.8b and c from the applied strain distribution of Figure 2.8a.

EXAMPLE 2.3   EUROCODE TEMPERATURE DISTRIBUTION IN SOLID CONCRETE BRIDGE

A 1000 mm deep solid concrete slab with 100 mm asphalt surfacing is subject to the Eurocode differential temperature distributions illustrated in Figure 2.9. Find the equivalent bending moments in terms of the coefficient of thermal expansion α and Young's modulus E.

For the heating-up distribution, the heights are $h_1 = 0.15$ m, $h_2 = 0.25$ m and $h_3 = 0.2$ m. Hence, the distribution is as illustrated in Figure 2.10a. With reference to the blocks in the temperature distribution, the components of moment about the centroid are calculated in Table 2.3 for a 1000 mm strip of slab. Note that the contribution from block d is negative – it is clockwise, whereas all the other contributions are anti-clockwise. The total moment is the sum of the contributions of each block and is equal to $521\alpha E$ kNm for the 1000 mm strip. If unrestrained, this positive moment will cause the centre of the slab to rise upwards.

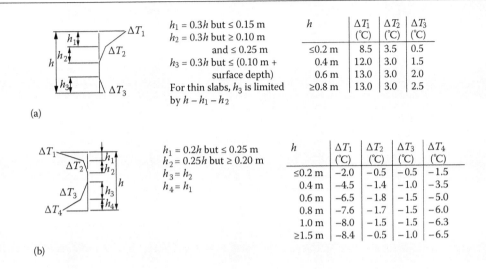

(a)

(b)

$h_1 = 0.3h$ but ≤ 0.15 m
$h_2 = 0.3h$ but ≥ 0.10 m
and ≤ 0.25 m
$h_3 = 0.3h$ but $\leq (0.10$ m +
surface depth)
For thin slabs, h_3 is limited
by $h - h_1 - h_2$

h	ΔT_1 (°C)	ΔT_2 (°C)	ΔT_3 (°C)
≤ 0.2 m	8.5	3.5	0.5
0.4 m	12.0	3.0	1.5
0.6 m	13.0	3.0	2.0
≥ 0.8 m	13.0	3.0	2.5

$h_1 = 0.2h$ but ≤ 0.25 m
$h_2 = 0.25h$ but ≥ 0.20 m
$h_3 = h_2$
$h_4 = h_1$

h	ΔT_1 (°C)	ΔT_2 (°C)	ΔT_3 (°C)	ΔT_4 (°C)
≤ 0.2 m	−2.0	−0.5	−0.5	−1.5
0.4 m	−4.5	−1.4	−1.0	−3.5
0.6 m	−6.5	−1.8	−1.5	−5.0
0.8 m	−7.6	−1.7	−1.5	−6.0
1.0 m	−8.0	−1.5	−1.5	−6.3
≥ 1.5 m	−8.4	−0.5	−1.0	−6.5

Figure 2.9 Eurocode rules for differential temperature distributions in concrete bridges: (a) heating up; (b) cooling down. (From British Standards Institution (BSI – www.bsigroup.com), Actions on structures, General actions, Thermal actions, BS EN 1991-1-5:2003 Eurocode 1, Extract reproduced with permission.)

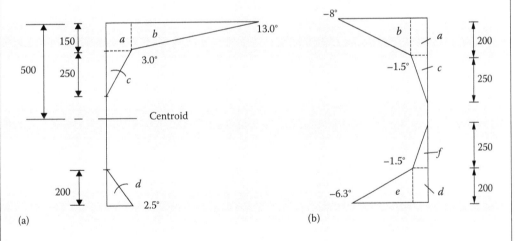

(a)

(b)

Figure 2.10 Differential temperature distributions for Example 2.3: (a) heating up; (b) cooling down.

Table 2.3 Calculation of heating-up moment for Example 2.3

Block	Details	Moment (N m)
a	$3\alpha E(1000 \times 150)(500 - 75) =$	$191 \times 10^6 \alpha E$
b	$\frac{1}{2}(10\alpha E)(1000 \times 150)(500 - 50) =$	$338 \times 10^6 \alpha E$
c	$\frac{1}{2}(3\alpha E)(1000 \times 250)(500 - 150 - 250/3) =$	$100 \times 10^6 \alpha E$
d	$-\frac{1}{2}(2.5\alpha E)(1000 \times 200)(500 - 200/3) =$	$-108 \times 10^6 \alpha E$

For the cooling down distribution, the heights are $h_1 = h_4 = 0.20$ m and $h_2 = h_3 = 0.25$ m, and the distribution is as illustrated in Figure 2.10b. The components of the equivalent moment due to temperature are given in Table 2.4, and the total moment is $-74\alpha E$ kNm/m.

Table 2.4 Calculation of cooling-down moment for Example 2.3

Block	Details	Moment (N m)
a	$-1.5\alpha E(1000 \times 200)(500 - 100) =$	$-120 \times 10^6 \alpha E$
b	$-\frac{1}{2}(6.5\alpha E)(1000 \times 200)(500 - 200/3) =$	$-282 \times 10^6 \alpha E$
c	$-\frac{1}{2}(1.5\alpha E)(1000 \times 250)(500 - 200 - 250/3) =$	$-41 \times 10^6 \alpha E$
d	$1.5\alpha E(1000 \times 200)(500 - 100) =$	$120 \times 10^6 \alpha E$
e	$\frac{1}{2}(4.8\alpha E)(1000 \times 200)(500 - 200/3) =$	$208 \times 10^6 \alpha E$
f	$\frac{1}{2}(1.5\alpha E)(1000 \times 250)(500 - 200 - 250/3) =$	$41 \times 10^6 \alpha E$

EXAMPLE 2.4   DIFFERENTIAL TEMPERATURE IN BEAM-AND-SLAB SECTION

For the beam-and-slab bridge illustrated in Figure 2.11a, the equivalent axial force, bending moment and residual stresses are required due to the differential temperature increases shown in Figure 2.11b. The coefficient of thermal expansion is α, and the modulus of elasticity is E.

By summing moments of area, the centroid of the bridge is found to be $\bar{x} = 0.318$ m below the top fibre. The bridge is split into two halves, each of area 0.70 m², and the second moment of area of each is 0.06486 m⁴. The temperature distribution is converted into a stress distribution in Figure 2.12 and divided into rectangular and triangular blocks. The total tensile force per half is then found by summing the products of stress and area for each block as shown in Table 2.5. The total force of $3.22\alpha E$ corresponds to an axial tension of $3.22\alpha E/0.70 = 4.60\alpha E$. Similarly,

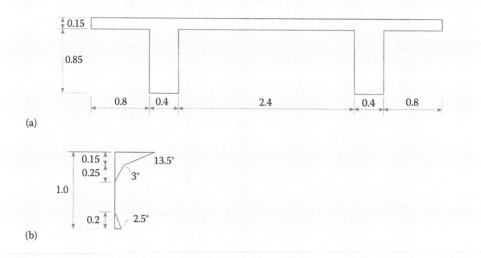

Figure 2.11 Beam-and-slab bridge subject to differential temperature: (a) cross section; (b) imposed temperature distribution (dimensions in metres).

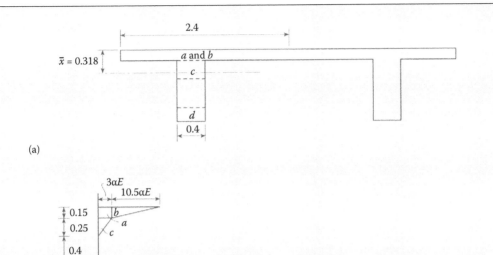

Figure 2.12 Division of section into blocks: (a) cross section; (b) corresponding imposed stress distribution.

Table 2.5 Calculation of force for Example 2.4

Block	Details	Force (kN)
a	$3\alpha E(2.4 \times 0.15) =$	$1.08\alpha E$
b	$\frac{1}{2}(10.5\alpha E)(2.4 \times 0.15) =$	$1.89\alpha E$
c	$\frac{1}{2}(3\alpha E)(0.4 \times 0.25) =$	$0.15\alpha E$
d	$\frac{1}{2}(2.5\alpha E)(0.4 \times 0.2) =$	$0.10\alpha E$

moment is calculated as the sum of products of stress, area and distance from the centroid as outlined in Table 2.6 (positive sag). The total moment of $-0.720\alpha E$ corresponds to stresses (positive tension) of

$$\text{stress on top} = -\frac{M\bar{x}}{I} = -\frac{(-0.720\alpha E)(0.318)}{0.06486} = 3.53\alpha E$$

Table 2.6 Calculation of moment for Example 2.4

Block	Details	Moment (kNm)
a	$-3\alpha E(2.4 \times 0.15)(\bar{x} - 0.15/2) =$	$-0.262\alpha E$
b	$-\frac{1}{2}(10.5\alpha E)(2.4 \times 0.15)(\bar{x} - 0.15/3) =$	$-0.507\alpha E$
c	$-\frac{1}{2}(3\alpha E)(0.4 \times 0.25)(\bar{x} - 0.15 - 0.25/3) =$	$-0.013\alpha E$
d	$\frac{1}{2}(2.5\alpha E)(0.4 \times 0.2)(1 - \bar{x} - 0.2/3) =$	$0.062\alpha E$

$$\text{stress on bottom} = \frac{M(1.0 - \bar{x})}{I} = \frac{(-0.720\alpha E)(1 - 0.318)}{0.06486} = -7.57\alpha E$$

Hence, the applied stress distribution can be resolved as illustrated in Figure 2.13. The residual distribution is found by subtracting the distributions of Figure 2.13b and c from the applied distribution of Figure 2.13a.

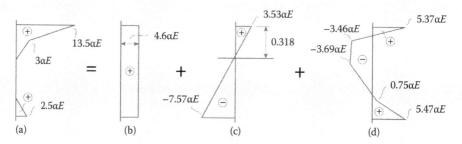

Figure 2.13 Resolution of stress distribution into axial, bending and residual components: (a) total distribution; (b) axial component; (c) bending component; (d) residual stress distribution.

2.6 IMPACT LOADING

Most bridge analysis is based on static linear elastic principles. However, the collision of a vehicle with a bridge is highly non-linear. The Eurocode describes two alternative simplifications: hard impact and soft impact. Hard impact is when an elastic vehicle hits an immovable structure, effectively 'bouncing' off it. Soft impact is when an incompressible vehicle hits an elastic–plastic structure, which absorbs the energy of the impact. Hard impact is considered here – this assumption allows the calculation of an equivalent static force for which the structure can be designed.

Hard impact is illustrated in Figure 2.14. A vehicle with mass, m, and stiffness, K, travelling at a velocity, v, collides with a structure of infinite stiffness. The kinetic energy of the vehicle is

$$E_k = \tfrac{1}{2} mv^2 \qquad (2.1)$$

On impact, this is converted into strain energy in the spring that represents the vehicle's elasticity. A static force, P_{eq}, which causes an elastic deflection in the vehicle, Δ, generates a strain energy of

$$\text{strain energy} = \tfrac{1}{2} P_{eq}\Delta \qquad (2.2)$$

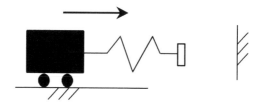

Figure 2.14 Impact of elastic vehicle with immovable structure.

Hence, the equivalent static force is

$$P_{eq} = \frac{2E_k}{\Delta} \tag{2.3}$$

For a spring of stiffness, K, a force P_{eq} generates an elastic deflection of

$$\Delta = P_{eq}/K \tag{2.4}$$

Substituting for Δ in Equation 2.3 gives an alternative expression for P_{eq}:

$$P_{eq} = \sqrt{2E_k K} \tag{2.5}$$

Substituting for E_k in this equation gives the equivalent force in terms of mass and velocity:

$$P_{eq} = v\sqrt{mK} \tag{2.6}$$

While this is a very simple case, it can be used as a basis for determining equivalent static forces. A table of design static forces is specified in the Eurocode based on the expected masses and velocities of trucks on roads of various classes. A simpler equivalent static load of 1780 kN (400 kip), 1.2 m (4 ft.) above the road surface, is specified in the AASHTO standard.

The Eurocode also specifies an impact force for a derailed train colliding with a pier. On bridges over road carriageways, there is a possibility that trucks passing underneath will collide with the bridge deck. However, because only the top of the vehicle is likely to impact on the bridge, a substantial reduction factor applies. It is not necessary, in the Eurocode, to consider collision of trains with bridge decks overhead.

2.7 DYNAMIC EFFECTS

Vibration can be a problem in slender bridges where the natural frequency is at a level that can be excited by traffic or wind. Vibration problems are most commonly encountered in long spans or in slender pedestrian bridges. However, there are also a number of regular bridges that, for one reason or another, have significant vibrations.

The vast majority of bridges do not have significant vibration problems but an allowance is made at the design stage for the dynamic interaction of vehicle and bridge. The greatest dynamic amplifications occur when there is an element of resonance, that is, when some frequency associated with the crossing of the vehicle excites a natural frequency of the bridge. Bridge frequency is a function of span, stiffness (second moment of area) and mass per unit length. A good rule of thumb is that the first natural frequency is usually approximately 100th of the span (Heywood et al. 2001).

Early work (Frýba 1971; Brady and OBrien 2006) considers single forces moving across beams. This suggests that there are a number of pseudo-frequencies associated with vehicle-crossing events. For example, the time it takes for the vehicle to travel from the start to the centre of a bridge can be considered to be a pseudo-period and its reciprocal a pseudo-frequency. If this pseudo-period matches half the first natural period of the bridge, then the force will reach the centre of the bridge at the same time that the bridge is at the extreme of its first vibration cycle, which will result in a peak of bending moment. Equally, if the pseudo-period matches a multiple of the first natural period, there will be another peak in bending moment. This is illustrated in Figure 2.15, which shows dynamic amplification factor (DAF) as a function of the ratio of vehicle speed to bridge first natural frequency. It shows that, at least in theory, increasing speed can either increase or decrease dynamic amplification but that there is an underlying increasing trend.

Brady et al. (2006) go on to show that field measurements do not reproduce the trend evident in Figure 2.15; when a truck crossed a bridge at a range of speeds, there was no obvious relationship between dynamic amplification and speed. There are several reasons for this: a real vehicle has mass and multiple axles and there are new pseudo-frequencies associated with the axle spacings. The vehicle also has a suspension system and tyres that behave like springs. Significantly, the road surface has a number of frequencies present in its profile that excite the vehicle. Cebon (1999) reports that vehicles have body bounce and pitch frequencies around 1.5–4 Hz and axle hop frequencies in the range of 8–15 Hz.

Field measurements by Žnidarič and Lavrič (2010) show that dynamic amplification tends to reduce as the weight of vehicles on the bridge increases (Figure 2.16), a finding supported by anecdotal evidence. It should be noted that the static weights in this graph are inferred from the dynamic measurements using the theory of bridge weigh-in-motion (OBrien et al. 1999; McNulty and OBrien 2003; González et al. 2008; Richardson et al. 2014), which introduces a risk of bias. However, OBrien et al. (2013) investigate this issue and suggest that the bias effect is not significant. Figure 2.16 shows that, while DAF can be as high as 2.0 in some cases, it is much less for extreme loading cases when the vehicles and vehicle combinations are much heavier. This decreasing trend as vehicle(s) get heavier is consistent with the codes of practice in some countries.

Caprani et al. (2012) and OBrien et al. (2009) suggest that DAF is an inappropriate measure of dynamics in bridges. DAF is the slope corresponding to any point in a graph of total

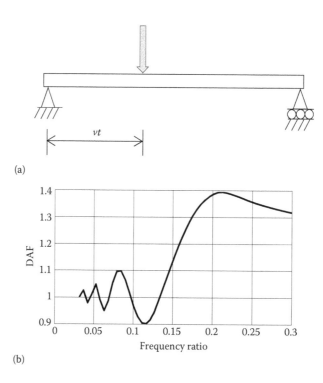

(a)

(b)

Figure 2.15 DAF for a moving force crossing a smooth simply supported beam: (a) geometry (*v* = vehicle speed, *t* = time); (b) DAF versus ratio of speed to first natural frequency.

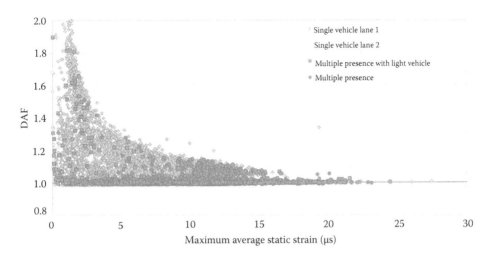

Figure 2.16 Dynamic amplification factor as a function of the strain due to the combined weight of all vehicles on a multi-span bridge in Slovenia. (From Žnidarič, A. and Lavrič, I., Applications of B-WIM technology to bridge assessment, *Bridge Maintenance, Safety, Management and Life-Cycle Optimization*, eds. D.M. Frangopol, R. Sause and C.S. Kusko, Philadelphia, Taylor & Francis, 1001–1008, 2010.)

(static + dynamic) versus static stress, as illustrated in Figure 2.17a. As such, it varies from one loading scenario to the next. It is inappropriate to take an extreme of DAF as it may correspond to a loading scenario where the stress is small. A more appropriate measure of dynamics is the assessment dynamic ratio (ADR), the ratio of characteristic total stress to characteristic static stress (Figure 2.17b). These characteristic values may correspond to different loading events or simply to a level of stress with an acceptably low probability of exceedance.

OBrien et al. (2010) report on an extensive study in which they statically simulate 10,000 years of traffic loading on two lanes for a range of bridge spans and LEs (bending moment and shear). For each of 10,000 maximum-per-year LEs, they carry out a vehicle/bridge dynamic interaction simulation. Typical bridge span/depth ratios are assumed, and vehicle properties and road surface profiles are varied randomly within typical ranges. Both the static and the total (static + dynamic) maximum-per-year data are plotted on probability paper in Figure 2.18 (see Section 2.3.2 for explanation of probability paper). The statistical trend for the maximum-per-year bending moments can be seen to be similar for static and

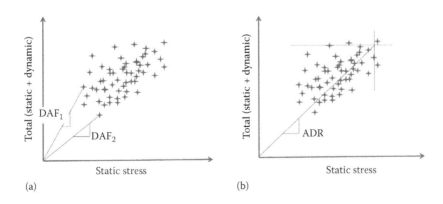

Figure 2.17 Allowances for dynamics: (a) DAF for different loading events; (b) ADR.

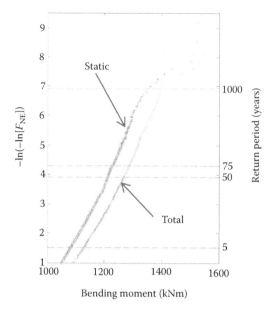

Figure 2.18 Probability of non-exceedance, F_{NE}, versus maximum-per-year bending moment in a 45 m span simply supported bridge.

total but with a small shift to the right for the total. The ADR is the ratio of characteristic total to characteristic static LE for any given return period and can be taken from graphs of this type. A summary of some results is presented in Figure 2.19. They find no clear trend in ADR with bridge span, but, for all LEs and spans considered, the ADR values are less than 1.1, that is, they find that the allowance for dynamics does not exceed 10%.

The Eurocode allowances for dynamics are included in the traffic load model of Figure 2.4a. The allowance varies from 30% to 10% for two-lane bridges, depending on span (Dawe 2003; Caprani et al. 2012). The AASHTO code specifies a separate allowance for dynamics

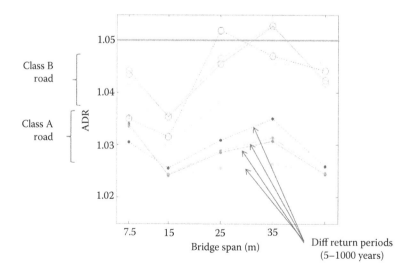

Figure 2.19 ADRs for mid-span moment for different bridge spans, return periods and road-surface roughnesses.

for the vehicle or tandem part of the HL93 load model of Figure 2.4b. Except for fatigue, this allowance is a fixed level of 33%, independent of span, number of lanes and LE type.

For railway bridges, the Eurocode specifies a dynamic factor for the LEs that result from the load model, LM71 or SW/2 of Section 2.3.4. The dynamic factor, Φ, varies as a function of the 'determinant length', L_Φ, which in turn depends on the structural element being considered and on whether or not ballast is present. The AREMA manual also specifies dynamic factors, which it calls impact factors. These are also a function of length and whether or not ballast is present. Different formulas are given for concrete and steel structures.

2.8 PRESTRESS LOADING

While prestress is not in fact a loading as much as a means of resisting load, it is often convenient to treat it as a loading for analysis purposes. Like temperature, prestress can be handled using the method of equivalent loads. However, it is important to note that prestress is an applied stress whereas temperature is an applied strain. Analysis for the effects of prestress is only necessary in the case of indeterminate bridges. However, even for simply supported slab or beam-and-slab bridges, it is often necessary to analyse to determine the degree to which prestressing of one member affects others. Whether the bridge consists of beams or a slab, equivalent loadings can be found for individual tendons. The combined effect of a number of tendons can then be found by simply combining the loadings and analysing.

Examples of analysis using equivalent prestress loads are given in Chapter 3. This section is focused on the calculation of the magnitudes of these loads. For a qualitative understanding of the effects of prestress, the concept of linear transformation is also introduced.

2.8.1 Equivalent loads and linear transformation

For the externally prestressed bridge illustrated in Figure 2.20a, the equivalent vertical loading at A is simply

$$F_A = P\sin\theta$$

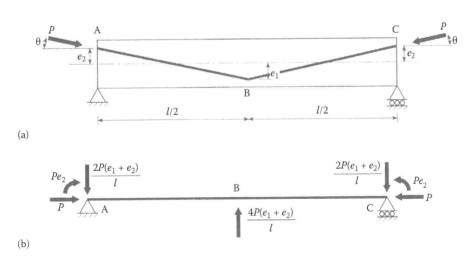

(a)

(b)

Figure 2.20 Prestressed concrete beam with external post-tensioning: (a) elevation showing tendon; (b) equivalent loading due to prestress.

As the angle, θ, is generally small, this can be approximated as

$$F_A \approx P\tan\theta = P(e_1 + e_2)/(l/2) = 2P(e_1 + e_2)/l \tag{2.7}$$

Other equivalent loadings due to prestress can be found by simple equilibrium of forces. For example, equilibrium of vertical forces gives an upward force at B of

$$F_B = 2P\sin\theta \approx 4P(e_1 + e_2)/l \tag{2.8}$$

It also follows from the small angle that the horizontal force is $P\cos\theta \approx P$. Finally, as the forces are eccentric to the centroid at the ends, there are concentrated moments there of magnitude $(P\cos\theta)e_2 \approx Pe_2$. Hence, the total equivalent loading due to prestress is as illustrated in Figure 2.20b.

It can be shown that the equivalent loading due to prestress is always self-equilibrating. A parabolically profiled prestressing tendon generates a uniform loading, which again can be quantified using equilibrium of vertical forces. A small segment of such a profile is illustrated in Figure 2.21a. At point 1, there is an upward vertical component of the prestress force of

$$F_1 = P\sin\theta_1 \tag{2.9}$$

As the angles are small

$$\sin\theta_1 \approx \tan\theta_1 = \left.\frac{dy}{dx}\right|_{x=x_1} \tag{2.10}$$

$$\Rightarrow \quad F_1 = P\left.\frac{dy}{dx}\right|_{x=x_1} \tag{2.11}$$

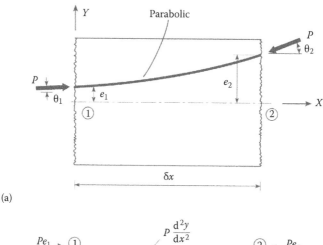

(a)

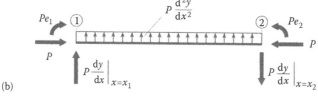

(b)

Figure 2.21 Segment of parabolically profiled tendon: (a) elevation; (b) equivalent loading.

where x_1 is the X coordinate at point 1. This force is upwards when the slope is positive. Similarly, the vertical component of force at 2 is

$$F_2 = P \left. \frac{dy}{dx} \right|_{x=x_2} \tag{2.12}$$

where F_2 is downwards when the slope is positive. The intensity of uniform loading on this segment is, by equilibrium,

$$w = \frac{F_2 + F_1}{\delta x}$$

$$= \frac{P \left(\left. \dfrac{dy}{dx} \right|_{x=x_2} - \left. \dfrac{dy}{dx} \right|_{x=x_1} \right)}{\delta x}$$

$$\Rightarrow w = P \frac{d^2 y}{dx^2} \tag{2.13}$$

All of the equivalent loads on the segment are illustrated in Figure 2.21b.

EXAMPLE 2.5   PARABOLIC PROFILE

The beam illustrated in Figure 2.22 is prestressed using a single parabolic tendon set out according to the equation

$$y = e_A + (e_B - e_A) \frac{x}{l} - 4s \left[\frac{x}{l} - \left(\frac{x}{l} \right)^2 \right] \tag{2.14}$$

where s is referred to as the sag in the tendon over length l as indicated in Figure 2.22. It is required to determine the equivalent loading due to prestress.

Differentiating Equation 2.14 gives

$$\frac{dy}{dx} = \frac{e_B - e_A}{l} - 4s \left(\frac{1}{l} - \frac{2x}{l^2} \right) \tag{2.15}$$

As θ_A is small

$$\sin\theta_A \approx \tan\theta_A = \left. \frac{dy}{dx} \right|_{x=0}$$

$$\Rightarrow \sin\theta_A \approx \frac{e_B - e_A - 4s}{l} \tag{2.16}$$

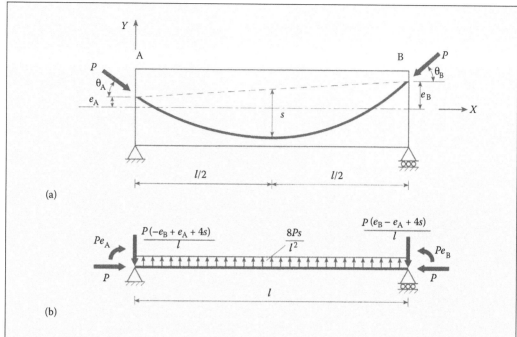

Figure 2.22 Beam with parabolic tendon profile: (a) elevation; (b) equivalent loading due to prestress.

For a positive slope, the equivalent point load at A would be upwards and of magnitude $P(e_B - e_A - 4s)/l$. However, in this case, the slope is negative and the force is downwards of magnitude $P(-e_B + e_A + 4s)/l$. The slope at B is calculated similarly:

$$\sin\theta_B \approx \left.\frac{dy}{dx}\right|_{x=l} = \frac{e_B - e_A + 4s}{l} \tag{2.17}$$

As B is on the right-hand side, this force is downwards when positive. Hence, the equivalent point loads are as illustrated in Figure 2.22b. The intensity of uniform loading is given by Equation 2.13 where the second derivative is found by differentiating Equation 2.14 twice:

$$\frac{d^2y}{dx^2} = \frac{8s}{l^2} \tag{2.18}$$

$$\Rightarrow \quad w = \frac{8Ps}{l^2} \tag{2.19}$$

This also is illustrated in the figure.

Example 2.5 illustrates the fact that the intensity of equivalent uniform loading due to a parabolic tendon profile is independent of the end eccentricities. A profile such as that illustrated in Figure 2.22a can be adjusted by changing the end eccentricities, e_A and e_B, while keeping the sag, s, unchanged. Such an adjustment is known as a linear transformation and will have no effect on the intensity of equivalent uniform loading as can be seen from Equation 2.19. This phenomenon is particularly useful for understanding the effect of prestressing in continuous beams with profiles that vary parabolically in each span.

EXAMPLE 2.6 QUALITATIVE PROFILE DESIGN

A prestressed concrete slab bridge is to be reinforced with 10 post-tensioned tendons. The preliminary profile for the tendons, illustrated in Figure 2.23a, results in insufficient compressive stress in the top fibres of the bridge at B. It is required to determine an amendment to the profile to increase the stress at this point without increasing the prestress force.

In a determinate structure, stress at the top fibre can be increased by moving the prestressing tendon upwards to increase the eccentricity locally. This increase in tendon eccentricity, e, increases the (sagging) moment due to prestress, Pe, which increases the compressive stress at the top fibre. However, in an indeterminate structure, the implication of such a change is not so readily predictable. In the structure of Figure 2.23, increasing the eccentricity locally at B without changing the sags, as illustrated in Figure 2.23b, does little to increase the compressive stress at the top fibre at that point. This is because the eccentricity at B has been increased without increasing the tendon sag in the spans. As was seen above, the equivalent uniform loading due to prestress is a function only of the sag and is, in fact, unaffected by eccentricity at the ends of the span. Thus, the change only results in adjustments to the equivalent point loads at A and B and to the equivalent loading near B. As these forces are at or near supports, they do not significantly affect the distribution of bending moment induced by prestress. A more appropriate revision is illustrated in Figure 2.23c where the profile is lowered in AB and BC while maintaining its position at the support

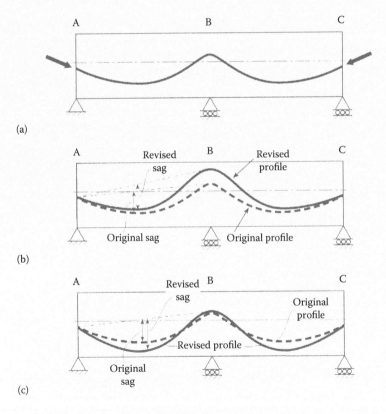

Figure 2.23 Adjustment of tendon profile: (a) original profile; (b) raising of profile at B by linear transformation; (c) lowering of profile in AB and BC to increase sag.

points. This has the effect of increasing the tendon sag, which increases the intensity of equivalent uniform loading. Such a uniform upward loading in a two-span beam generates sagging moment at the interior support, which has the desired effect of increasing the top-fibre stress there.

Most prestressing tendons are made up of a series of lines and parabolas, and the equivalent loading consists of a series of point forces and segments of uniform loading. This can be seen in the following example.

EXAMPLE 2.7 TENDON WITH CONSTANT PRESTRESS FORCE

A three-span bridge is post-tensioned using a five-parabola symmetrical profile, half of which is illustrated in Figure 2.24a. It is required to determine the equivalent loading due to prestress, assuming that the prestress force is constant throughout the length of the bridge. The intensities of loading are found from Equation 2.13. For the first parabola

$$y = -0.1322x + 0.01135x^2$$

$$\Rightarrow \frac{d^2y}{dx^2} = 0.0227$$

$$\Rightarrow w_{AC} = 0.0227P$$

Similarly, the intensities of loading in the second and third parabolas are, respectively,

$$w_{CD} = -0.05434P$$

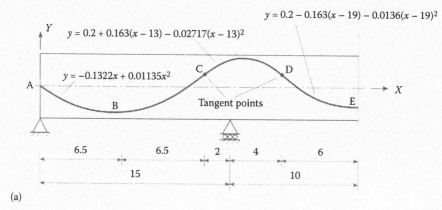

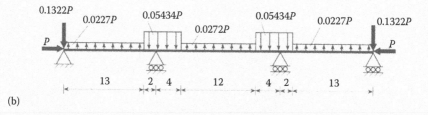

Figure 2.24 Tendon profile for Example 2.7: (a) partial elevation showing segments of parabola; (b) equivalent loading due to prestress.

and

$w_{DE} = 0.0272P$

The point load at the end support is the vertical component of the prestress force. Differentiating the equation for the parabola gives the slope, from which the force is found to be

$F_A = P \sin\theta_A \approx P \tan\theta_A = -0.1322P$

All of the equivalent loads due to prestress are illustrated in Figure 2.24b. Verifying that these forces are in equilibrium can be a useful check on the computations. Note that in selecting the profile, it has been ensured that the parabolas are tangent to one another at the points where they meet. This is necessary to ensure that the tendon does not generate concentrated forces at these points.

2.8.2 Prestress losses

In practical post-tensioned construction, prestress forces are not constant through the length of bridges because of friction losses. This is illustrated in Figure 2.25a where the forces at points 1 and 2 are different.

However, the difference between prestress forces at adjacent points is generally not very large. Therefore, a sensible approach to the derivation of equivalent prestress loading is to start by substituting the average prestress force for P in Equations 2.11–2.13. The resulting loading is illustrated in Figure 2.25b. It will be seen in Example 2.8 that this equivalent loading satisfies equilibrium of forces and moments. The use of equivalent loads that do not satisfy equilibrium can result in significant errors in the calculated distribution of prestress moment. A useful method of checking the equivalent loads is to apply them in the analysis of a determinate beam. In such a case, the moment due to the equivalent loading should be equal to the product of prestress force and eccentricity at all points.

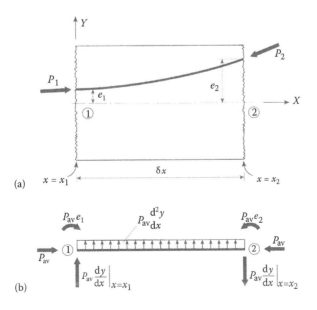

Figure 2.25 Equivalent loading due to varying prestress force: (a) segment of beam and tendon; (b) equivalent loading ($P_{av} = (P_1 + P_2)/2$).

EXAMPLE 2.8   TENDON WITH VARYING PRESTRESS FORCE

The post-tensioning tendon of Example 2.7 is subject to friction losses, which result in the pre-stress forces presented in Figure 2.26. The eccentricities given in this figure have been calculated from the equations for each parabola given in Example 2.7. It is required to determine the equivalent loading due to prestress, taking account of the loss of force. The bridge is post-tensioned from both ends, with the result that the prestressing forces vary symmetrically about the centre.

With reference to Example 2.7 but using average prestress forces, the equivalent intensities of uniform loading are

$$w_{AB} = 0.0227\left(\frac{5000+4800}{2}\right) = 111.2 \text{ kN/m}$$

$$w_{BC} = 0.0227\left(\frac{4800+4550}{2}\right) = 106.1 \text{ kN/m}$$

$$w_{CD} = -0.05434\left(\frac{4550+4350}{2}\right) = -241.8 \text{ kN/m}$$

$$w_{DE} = 0.0272\left(\frac{4350+4150}{2}\right) = 115.6 \text{ kN/m}$$

In addition, point loads must be applied at the end of each segment in accordance with Figure 2.25b. In segment AB, the equation for the parabola is

$$y = -0.1322x + 0.01135x^2$$

$$\Rightarrow \quad \frac{dy}{dx} = -0.1322 + 0.0227x$$

At A, $x = 0$, the slope is -0.1322 and the upward force is

$$P_{av}\frac{dy}{dx}\bigg|_{x=0} = \left(\frac{5000+4800}{2}\right)(-0.1322)$$

$$= -647.8 \text{ kN}$$

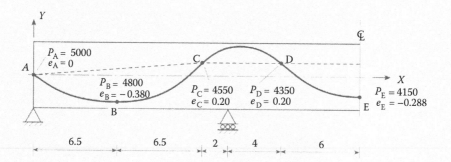

Figure 2.26 Tendon profile showing varying prestress force (in kN) and eccentricity (in m).

the minus sign indicating that the force is actually downwards. At B, the slope of the profile is

$$\left.\frac{dy}{dx}\right|_{x=6.5} = 0.1535$$

giving a downward force at the right end of magnitude

$$\left(\frac{5000 + 4800}{2}\right)(0.01535) = 75.2 \text{ kN}$$

The corresponding point load components for the other segments of parabola are calculated similarly and are presented, together with the other equivalent uniform loads, in Figure 2.27a. It can be verified that the forces and moments on each segment are in equilibrium. The forces and moments at the ends of each segment are summed, and the result is illustrated in Figure 2.27b.

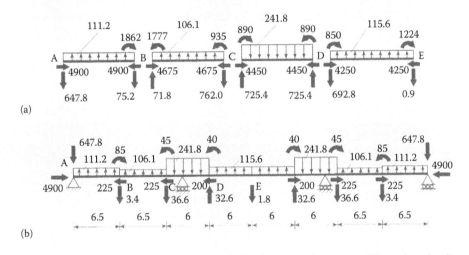

Figure 2.27 Equivalent loading due to prestress: (a) loading on each segment; (b) total net loading.

2.8.3 Non-prismatic bridges

The eccentricity of a prestressing tendon is measured relative to the section centroid. In non-prismatic bridge decks, the location of this centroid varies along the length of the bridge. This clearly affects the eccentricity and hence the moment due to prestress. A segment of beam with a curved centroid is illustrated in Figure 2.28a. In such a beam, the prestress forces are resolved parallel and perpendicular to the centroid, and the eccentricity is measured in a direction perpendicular to it. The resulting equivalent loading is illustrated in Figure 2.28b where *s* is the distance along the centroid.

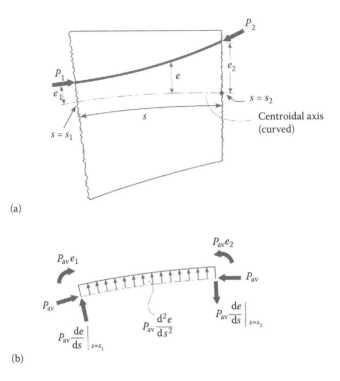

(a)

(b)

Figure 2.28 Equivalent loading due to variation in location of centroid: (a) segment of beam and tendon; (b) equivalent loading.

EXAMPLE 2.9   EQUIVALENT LOADING DUE TO CHANGE IN GEOMETRY

The beam illustrated in Figure 2.29 has a non-prismatic section – the centroid changes depth linearly between A and B, and between B and C. It is prestressed with a tendon, following a single parabolic profile from A to C. In addition, there are friction losses of 12%, which vary linearly between A and C (friction losses generally do not vary linearly, but this is a widely accepted approximation). It is required to determine the equivalent loading due to prestress. The beam is divided into just two segments: AB and BC. The definition of the parabola is independent of

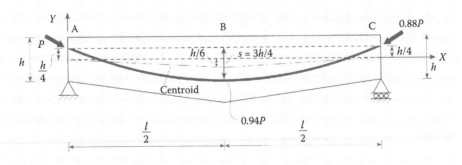

Figure 2.29 Elevation of beam and tendon profile.

the section geometry. With reference to Example 2.5, it is defined by an equation of the same form as Equation 2.14, that is,

$$y = e_A + (e_C - e_A)\frac{x}{l} - 4s\left[\frac{x}{l} - \left(\frac{x}{l}\right)^2\right]$$

$$= \frac{h}{4} - 4\left(\frac{3h}{4}\right)\left[\frac{x}{l} - \left(\frac{x}{l}\right)^2\right]$$

$$= \frac{h}{4} - 3h\left(\frac{x}{l}\right) + 3h\left(\frac{x}{l}\right)^2$$

If the eccentricity is approximated as the vertical distance, generally an excellent approximation, it can be found as the difference between y and the line representing the centroid. Hence, for segment AB

$$e = \frac{h}{4} - 3h\left(\frac{x}{l}\right) + 3h\left(\frac{x}{l}\right)^2 - \left(-\frac{h}{6}\frac{x}{(l/2)}\right)$$

$$= \frac{h}{4} - \frac{8h}{3}\left(\frac{x}{l}\right) + 3h\left(\frac{x}{l}\right)^2$$

Similarly for segment BC, the eccentricity is given by

$$e = \frac{5h}{12} - \frac{10h}{3}\left(\frac{x}{l}\right) + 3h\left(\frac{x}{l}\right)^2$$

Differentiating the equation for segment AB gives

$$\left.\frac{de}{ds}\right|_{AB} \approx \left.\frac{de}{dx}\right|_{AB} = -\frac{8h}{3l} + \frac{6hx}{l^2}$$

$$\Rightarrow \left.\frac{de}{ds}\right|_A = -\frac{8h}{3l}; \left.\frac{de}{ds}\right|_B = \frac{h}{3l}$$

Similarly for BC, the derivatives are

$$\left.\frac{de}{ds}\right|_{BC} \approx \left.\frac{de}{dx}\right|_{BC} = -\frac{10h}{3l} + \frac{6hx}{l^2}$$

$$\Rightarrow \left.\frac{de}{ds}\right|_B = -\frac{h}{3l}; \left.\frac{de}{ds}\right|_C = \frac{8h}{3l}$$

Differentiating again gives, for both segments,

$$\frac{d^2e}{ds^2} = \frac{6h}{l^2}$$

The average values for prestress force in segments AB and BC are 0.97P and 0.91P, respectively, where P is the jacking force. The resulting equivalent loading due to prestress is illustrated for each segment in Figure 2.30a. The forces are combined in Figure 2.30b.

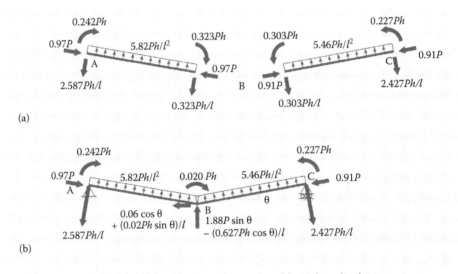

(a)

(b)

Figure 2.30 Equivalent loading: (a) loading on each segment; (b) total net loading.

Chapter 3

Introduction to bridge analysis

3.1 INTRODUCTION

This chapter considers the analysis of bridges where the transverse width is sufficiently small that they can be treated as beams or frames. This greatly simplifies the analysis so this chapter serves as an introduction to a range of analysis topics. Where transverse width is significant, a one-dimensional beam becomes a two-dimensional slab and a two-dimensional frame becomes a three-dimensional structure; these more complex structures are treated, for the most part, in later chapters.

Influence lines are shown to be useful in finding the critical locations of notional traffic load models. The method of equivalent loads is also presented as a means of analysing for the effects of phenomena such as temperature, prestress and creep.

3.2 POSITIONING THE TRAFFIC LOAD MODEL ON THE BRIDGE

Both the Eurocode Load Model 1 and the AASHTO HL93 notional load models need to be positioned on the bridge at the worst locations for each bending moment and shear force considered. This is commonly done using influence lines. An influence line is simply a graph of bending moment or shear force plotted against the location of a unit point force on the bridge. While influence lines can be derived from a series of analyses, with load cases for each possible location of the force, it is easier in the earlier stages of the design process to sketch them directly using the Maxwell–Betti theorem.

For bending moment, the Maxwell–Betti theorem involves inserting a pin at the point of interest and imposing a unit rotation. Surprisingly, the resulting shape is the influence line for moment at that point. Figure 3.1 illustrates the concept. The influence line for moment at the centre of the beam – point B – is found by inserting a pin at B and imposing a unit rotation there. As the beam is pinned at A and C, the only way there can be a rotation at B is if the beam lifts up at that point. The rotations at A and C are ½ because of symmetry and the fact that the unit external angle must equal the sum of the two internal angles. Hence, the magnitude of the displacement at B is *angle* × *distance* = (½ × $l/2$) = $l/4$.

For shear force, a vertical shift is imposed at the point of interest that maintains the slope on either side. The Maxwell–Betti rules are summarised in Figure 3.2.

Once the influence line is available, the moment or shear force is found by summing the applied forces and the corresponding influence line ordinates. For a uniformly distributed load, the moment or shear force is the product of the intensity and the area under the influence line. The concept is illustrated in the following example.

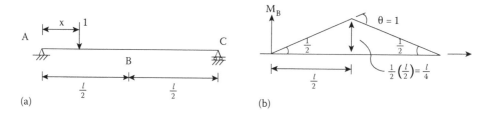

Figure 3.1 Finding the influence line for moment in a simply supported beam: (a) beam with unit load; (b) influence line for moment at B.

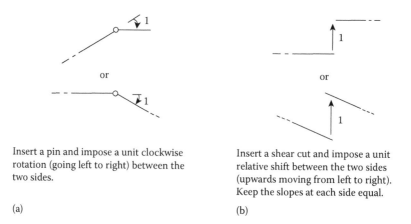

Insert a pin and impose a unit clockwise rotation (going left to right) between the two sides.

(a)

Insert a shear cut and impose a unit relative shift between the two sides (upwards moving from left to right). Keep the slopes at each side equal.

(b)

Figure 3.2 Summary of Maxwell–Betti rules: (a) internal bending moment; (b) shear force.

EXAMPLE 3.1   POSITIONING THE EUROCODE LM1 ON A SIMPLY SUPPORTED BRIDGE

A bridge is 20 m long and simply supported. The transverse width between kerbs is 8 m. Find the critical locations for the Eurocode LM1 load model for bending moment and shear force at the quarter, half and three-quarter points.

As the bridge is less than 9 m wide, there are only two notional lanes. Referring to Figure 2.4, the uniformly distributed loading is 9 kN/m² in lane 1 and 2.5 kN/m² in both lane 2 and the remaining 2 m width. This gives a total uniform loading of (9 × 3 + 2.5 × 5 =) 39.5 kN/m². The tandems in lanes 1 and 2 apply 600 and 400 kN, respectively, giving a total of 1000 kN.

The influence lines for moment and shear are given in Figure 3.3. The critical locations of the uniformly distributed loading and the bogie are shown in Figure 3.4. There are a few points of note in the latter figure:

- The bogie is positioned right of point B in Figure 3.4a as the influence ordinate falls off more rapidly left of B than right of it.
- For moment at point C, the bogie can be positioned left or right of the centre.
- For shear at B, C and D, the uniform loading is only applied to part of the bridge.
- It does not matter whether shear is positive or negative – the position that gives maximum absolute value of shear is chosen in Figures 3.4d–h.

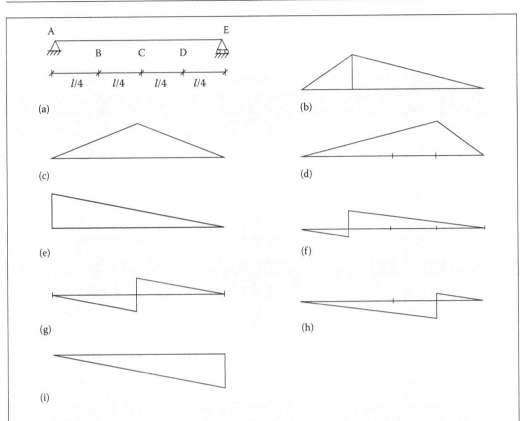

Figure 3.3 Beam of Example 3.1 and influence lines: (a) beam; (b) influence line for moment at B; (c) moment at C; (d) moment at D; (e) shear at A; (f) shear at B; (g) shear at C; (h) shear at D; (i) shear at E.

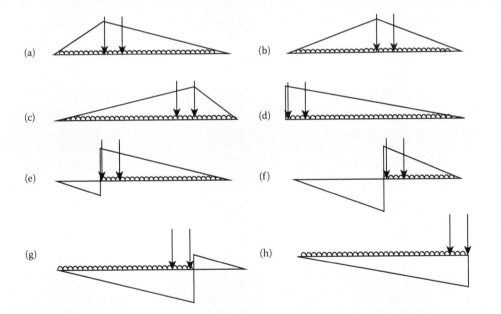

Figure 3.4 Positioning of LM1 for beam of Example 3.1: (a) moment at B; (b) moment at C; (c) moment at D; (d) shear at A; (e) shear at B; (f) shear at C; (g) shear at D; (h) shear at E.

EXAMPLE 3.2   POSITIONING THE HL93 LOAD MODEL ON AN INDETERMINATE BRIDGE

The bridge of Figure 3.5a is continuous with two equal spans of 30 m. The transverse width between kerbs is 13 m. Find the critical location for the HL93 load model for shear just left of the support at B.

According to the AASHTO code, the number of notional lanes is 13/3.66 = 3.55, which rounds down to 3. Allowing for the multiple presence factor of 0.85, HL93 on three notional lanes gives a total of (3 × 0.85 =) 2.55 load models. Hence, the uniform loading is (2.55 × 9.34 =) 23.8 kN/m. The HL93 truck (Figure 2.4b) has loads of 36, 142 and 142 kN for axles 1, 2 and 3, respectively. Allowing for 2.55 load models over the three lanes and the dynamic allowance of 1.33, these become 122, 482 and 482 kN.

For determinate structures, influence lines are made up of straight line segments. For indeterminate structures, the same principles of Figure 3.2 apply, but the segments are no longer necessarily straight lines. For the bridge of Figure 3.5a, the influence line for shear just left of B is illustrated in Figure 3.5b. The function must pass through zero at B so it must be at −1 just before that point. Both AB and BC are curved to satisfy the requirement that the slope just left of B equals the slope just right of it.

The critical position of the HL93 model is also shown in Figure 3.5b. The position of the truck can be tested for criticality by considering the implications of moving it to the left or the right. If it is moved left, the influence ordinates corresponding to all three axle loads reduce. If it is moved right, the influence ordinate corresponding to the first and last axles increase, but this is countered by a step-change reduction in the ordinate corresponding to the middle axle. The spacing between the first and second (rightmost) axles can vary in the range 4.27–9.14 m. The maximum value is chosen as it gives the greatest shear force.

The tandem axle is specified as an alternative that must be considered to the truck. Its two axles are lighter than axles of the truck but are more closely spaced. It follows that the tandem only governs when the third axle of the truck is reducing the moment or shear. For this example, all three axles are contributing to the shear force, so the tandem will not govern.

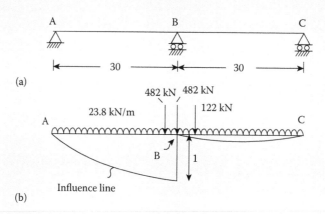

Figure 3.5 Two-span bridge of Example 3.2: (a) geometry; (b) influence line and critical location of HL93 truck model.

3.3 DIFFERENTIAL SETTLEMENT OF SUPPORTS

There is considerable research and development activity currently taking place in the field of soil/structure interaction. Clearly, soil deforms under the vertical forces applied through bridge piers and abutments. If the deformation is not uniform, distributions of bending moment and shear are induced in the deck. To accurately analyse for this effect, the structure and the surrounding soil may be represented using non-linear computer models. However, as the effect is often not very significant, some structural engineers treat the soil as a spring or a series of springs in the numerical model. The disadvantage of this is that differential settlement is more often caused by a relatively weak patch of soil under one support rather than by a non-uniform distribution of applied loads. Thus, an alternative approach, frequently adopted by bridge engineers, is to assume that a foundation support settles by a specified amount, Δ, relative to the others and to determine the effects of this on the structure. The following example serves to demonstrate the effect of a differential settlement on a continuous beam bridge.

EXAMPLE 3.3   DIFFERENTIAL SETTLEMENT OF SUPPORTS

The continuous beam illustrated in Figure 3.6 is subjected to a settlement at B of Δ relative to the other supports. The resulting bending-moment diagram (BMD) is required, given that the beam has uniform flexural rigidity, EI.

Imposing settlements at supports is an option in many of today's computer programs. Even if it is not an explicit option, it is usually possible to specify a spring support at B, to apply a vertical force to the spring and to scale that force to give the required settlement, Δ. The beam has been analysed here using the stiffness method and gives the BMD illustrated in Figure 3.7.

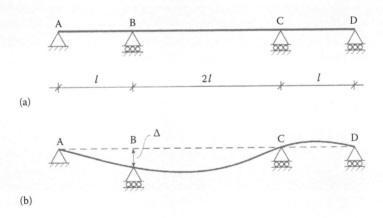

Figure 3.6 Three-span beam example: (a) geometry; (b) imposed support settlement.

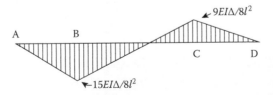

Figure 3.7 BMD due to imposed settlement of Δ at B.

Differential settlement has the effect of generating sagging moment at the support that settles. This is important as supports in continuous beams are generally subjected to hogging moment and are often not designed to resist significant sag. It is interesting to note two additional things about this BMD, which are typical of differential settlement:

1. The moment at the support that settles is proportional to the second moment of area, I, divided by the square of the span length, l. Hence, in general terms, differential settlement moments increase with cross-sectional stiffness, I, and decrease with span, l. Of course, longer bridges need to be stiffer, which complicates the issue.
2. Unlike BMDs due to applied forces, the distribution of moment due to differential settlement is proportional to the elastic modulus. This is particularly significant for concrete bridges where considerable creep occurs. A widely accepted approximate way to model the effect of creep is to reduce the elastic modulus. As moment is proportional to this modulus, it follows that creep has the effect of reducing the moment due to differential settlement over time. This beneficial effect of the creep in concrete is countered by the fact that the magnitude of the differential settlement itself often increases with time due to time-dependent behaviour in the supporting soil. However, if the specified settlement is deemed to include such time-dependent effects, it is reasonable to anticipate some reduction in moment due to concrete creep.

3.4 THERMAL EXPANSION AND CONTRACTION

As discussed in Chapter 2, there are two thermal effects for which bridge analysis is required, namely, axial expansion/contraction and differential changes in temperature through the depth of the bridge deck. In this section, analysis for the effects of axial expansion/contraction due to temperature changes is considered.

If a beam is on a sliding bearing, as illustrated in Figure 3.8a and the temperature is reduced by ΔT, it will contract freely. A (negative) strain will occur of magnitude $\alpha(\Delta T)$, where α is the coefficient of thermal expansion (strain per unit change in temperature). The beam then contracts by $\alpha(\Delta T)l$, where l is its length. However, no stresses are generated as no restraint is offered to the contraction. As there is no stress, there can be no tendency to crack. If, on the other hand, the beam is fixed at both ends, as illustrated in Figure 3.8b, and its temperature is reduced by ΔT, then there will be no strain – there cannot be any strain as the beam is totally restrained against contraction. This total restraint generates a stress of magnitude $E\alpha(\Delta T)$, where E is the elastic modulus. The stress is manifested in a tendency to crack.

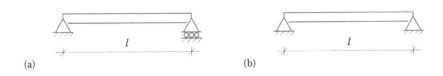

Figure 3.8 Extreme restraint conditions for axial temperature: (a) free; (b) fully fixed.

The most common case requiring analysis is the one in between the two extreme cases described above, where a beam is partially restrained. This happens, for example, in arch bridges where contraction is accommodated through bending in the arch (Figure 1.30). It also happens in frame bridges where the piers offer some resistance to expansion or contraction of the deck.

EXAMPLE 3.4   RESTRAINED AXIAL EXPANSION

The structure illustrated in Figure 3.9a is not typical of a real bridge but is included here to illustrate the phenomenon of restrained thermal expansion. It is pinned at B to a pier BD with stiffness EI. It is required to find the bending moment, shear force and axial force diagrams due to an increase in deck temperature of ΔT.

The beam is pinned at A, so expansion of AB is resisted by the pier, which must bend to accommodate it, as illustrated in Figure 3.9b. The coefficient of thermal expansion, α, is the strain per degree change in temperature. Hence, $\alpha(\Delta T)$ is the 'applied' strain in the deck. Because of the sliding bearing at C, the section of the beam right of B is fully free to expand and BC will increase its length by $4h\alpha\Delta T$.

Left of C, the same applied strain of $\alpha\Delta T$, is resisted by the pier, so it generates a slightly smaller increase in length (i.e. slightly less strain) and a little stress. The stiffnesses of the deck in axial deformation and the pier in bending are very different: the deck has stiffness $(area)E/$ $(length) = (6000I/h^2)E/(4h) = 1500EI/h^3$; the pier has stiffness $3EI/h^3$. It follows from the principle of relative stiffness that $(3/1503 =) 0.002$ of the applied strain causes stress in AB and bending moment in BD, while the remaining 0.998 causes movement at B. The full results of the analysis are illustrated in Figure 3.10.

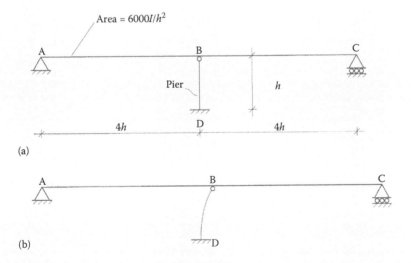

(a)

(b)

Figure 3.9 Frame subjected to axial change in temperature: (a) original geometry; (b) deformed shape after expansion of deck.

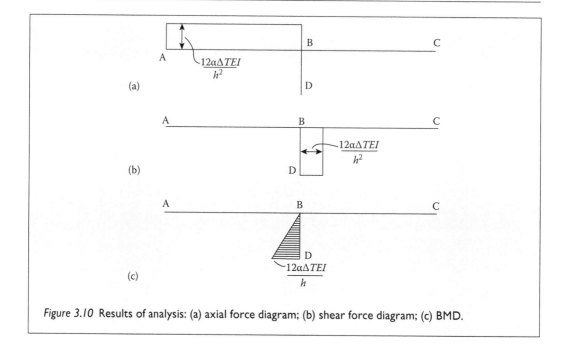

Figure 3.10 Results of analysis: (a) axial force diagram; (b) shear force diagram; (c) BMD.

There are some points of interest about axial temperature effects apparent from this simple example. What is most noteworthy is the effect of the relative values of deck area and pier second moment of area. The area of the deck is generally large and its stiffness generally much greater than the stiffness of any restraining piers, with the result that the restraint to deck expansion is relatively small. Hence, the rise in temperature results in a lot of strain and very little stress in the deck. It is also of interest to note that, as for differential settlement, the moments and forces due to changes in temperature are proportional to the elastic modulus. This means that such stresses, if sustained in a concrete structure, may be relieved by the effect of creep.

Substantial temperature changes occur on a short-term basis during which the effects of creep do not have a significant ameliorating effect. However, in situ concrete bridges generate significant quantities of heat while setting and consequently have their initial set when the concrete is warm. The sustained stresses generated by the subsequent contraction of the concrete as it cools may be relieved substantially by creep.

EXAMPLE 3.5   THERMAL CONTRACTION IN FRAME BRIDGE

The frame structure illustrated in Figure 3.11 is integral, having no internal bearings or joints. As a result, thermal contraction or expansion induces bending moment and shear as well as axial force. It is subjected to a uniform reduction in temperature through the depth of the deck (ABC) of 20°C and no change in temperature elsewhere. The resulting distribution of bending moment is required given that the coefficient of thermal expansion is 12×10^{-6}. The relative flexural rigidities are given in Figure 3.11, and the area of the deck is $500 I_0/l^2$.

Analysis of the frame bridge gives the results illustrated in Figure 3.12. The small axial force in the deck is balanced by shear forces in the outer piers. This example serves to illustrate the effect of a moment connection between the bridge deck and the piers. The applied thermal

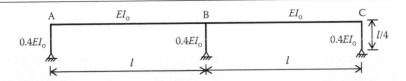

Figure 3.11 Integral frame of Example 3.5.

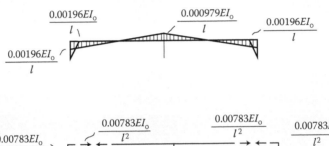

(a)

(b)

Figure 3.12 Results of analysis for effects of thermal contraction: (a) bending-moment diagram; (b) free body diagram.

movement is resisted by bending in both the piers and the deck. To some extent, this alters the resistance to contraction or expansion. However, a more important effect of the moment connection is the bending moment induced in the deck. This can become a significant factor in bridge deck design.

3.4.1 Equivalent loads method

The method of equivalent loads is a method by which a thermal expansion/contraction problem can be converted into a regular analysis problem. This approach is particularly useful when a computer is available to carry out the analysis, but the program does not cater directly for temperature effects.

EXAMPLE 3.6   INTRODUCTION TO EQUIVALENT LOADS METHOD

The equivalent loads method will first be applied to the simple problem of the partially restrained beam illustrated in Figure 3.13, which is subjected to an axial increase in temperature of ΔT. The expansion is partially restrained by a spring of stiffness $AE/(2l)$, where A is the cross-sectional area, and E is the elastic modulus of the beam. The equivalent loads method consists of three stages as follows:

Stage A – Calculate the equivalent loads and the associated stresses: The loading is found, which would generate the same strain in an unrestrained member as the distribution of temperature.

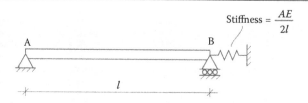

Figure 3.13 Beam on rollers with partial (spring) restraint.

An axial expansion can be generated in an unrestrained beam by applying an axial force, F_0, where

$$F_0 = A(\text{stress}) = AE\alpha\Delta T$$

where α is the coefficient of thermal expansion. However, temperature on an unrestrained member applies strain but not stress. The equivalent force, on the other hand, will apply both, even on an unrestrained beam. Therefore, it is necessary to identify the 'associated stresses', that is, that distribution of stress, which is inadvertently introduced into the structure by the equivalent loads. In stage C, this distribution of stress must be subtracted to determine the stresses generated indirectly by the change in temperature. The equivalent loads for this example are illustrated in Figure 3.14a and the associated stress distribution in Figure 3.14b.

Stage B – Analyse for the effects of the equivalent loads: The beam is analysed for the loading illustrated in Figure 3.14a. Normally, this stage would be done by computer, but it is trivial for

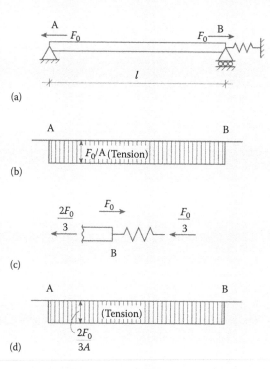

(a)

(b)

(c)

(d)

Figure 3.14 Analysis by equivalent-loads method: (a) equivalent loads; (b) associated stress distribution; (c) equilibrium of forces at spring; (d) stress distribution due to temperature change.

this simple example. It follows from the principle of relative stiffness that when a load is applied to two springs, it is resisted in proportion to their stiffnesses. In this case, the beam acts as a spring of stiffness AE/l. Hence, the force is taken in the ratio 1:2 as illustrated in Figure 3.14c. The distribution of stress due to application of F_0 is an axial tension throughout the beam of magnitude $2F_0/(3A)$, as illustrated in Figure 3.14d.

Stage C – Subtract the associated stresses: The distribution of associated stresses is subtracted from the stresses generated by the equivalent loads. For this example, this consists of subtracting the axial stress distribution of Figure 3.14b from that of Figure 3.14d. The result is an axial compression of $F_0/(3A)$ throughout the beam. This is the final result and is what one would expect from a thermal expansion in a partially restrained beam; not only strain is generated but also some compressive stress.

3.5 DIFFERENTIAL TEMPERATURE EFFECTS

When the sun shines on the top of a bridge, the top tends to increase in temperature faster than the bottom. Thus, a differential temperature distribution develops, which tends to cause the bridge to bend. If a linear distribution of this type is applied to a simply supported single-span beam, the bending takes place freely, and the beam curves upwards as the top expands relative to the bottom. This corresponds to the case of a beam on rollers subjected to an axial increase in temperature in that strains take place but not stresses. The essential difference is that axial temperature changes cause axial stress distributions (forces) and strains/deflections, whereas differential temperature changes cause bending stress distributions (moments) and curvatures/rotations. If such a differential temperature distribution is applied to a beam in which the ends are fixed against rotation, the free bending is prevented from taking place, and the situation is one of stress but no strain. In multi-span beams and slabs, partial restraint against bending is present as will be seen in the following examples.

EXAMPLE 3.7   DIFFERENTIAL TEMPERATURE IN A TWO-SPAN BEAM

The two-span beam illustrated in Figure 3.15 is subjected to a change of temperature, which is non-uniform through its depth. The temperature change varies linearly from an increase of 5° at the top to a decrease of 5° at the bottom. The centroid of the beam is at mid-height, the elastic modulus is E and the second moment of area is I. It is required to find the BMD due to the temperature change given that the coefficient of thermal expansion is α.

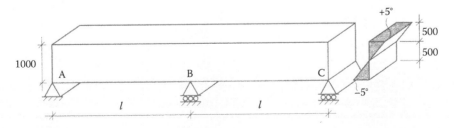

Figure 3.15 Beam of Example 3.7 and applied distribution of temperature.

The BMD will be found using the method of equivalent loads.

Stage A – Calculate the equivalent loads and the associated stresses: If unrestrained, the temperature change would generate a distribution of strain varying from 5α at the top to -5α at the bottom, where α is the coefficient of thermal expansion. Consider the familiar flexure formula:

$$\frac{M}{I} = \frac{E}{R} = \frac{\sigma}{y}$$

where M is the moment, R is the radius of curvature, σ is the stress and y is the distance from the centroid.

$$\Rightarrow \quad \frac{M}{EI} = \frac{1}{R} = \frac{\varepsilon}{y}$$

where ε is the strain. The ratio $1/R$ is known as the curvature, κ. In this case, the change in temperature generates a curvature of

$$\kappa = \varepsilon/y = 5\alpha/500 = \alpha/100 \text{ mm}^{-1}$$

The corresponding equivalent moment is

$$M = EI\kappa = EI\alpha/100 \text{ Nmm}$$

Temperature on an unrestrained structure generates strain and curvature but not bending moment or stress. The equivalent moment, on the other hand, will generate both curvature and bending moment, even on unrestrained beams. Therefore, it is necessary to identify the 'associated BMD', that is, that distribution of moment, which is inadvertently introduced into the structure by the equivalent loading. The equivalent loads and associated BMD are illustrated in Figure 3.16a and b, respectively.

Stage B – Analyse for the effects of the equivalent loads: Analysis of a symmetrical two-span beam is trivial because, due to symmetry, the central support point B does not rotate. Hence, it is effectively fixed, as illustrated in Figure 3.17a, and the solution can be determined directly from Appendix A. The BMD due to the applied equivalent loading is as illustrated in Figure 3.17b.

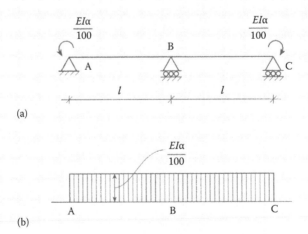

(a)

(b)

Figure 3.16 Application of equivalent loads method: (a) equivalent loads; (b) associated BMD.

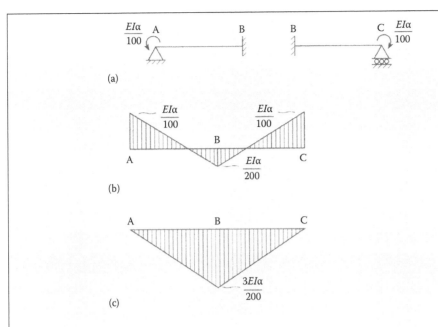

Figure 3.17 Stages in equivalent loads method: (a) applied equivalent loads; (b) BMD due to application of equivalent loads; (c) BMD after subtraction of associated BMD.

Stage C – Subtract the associated stresses: Subtracting the associated BMD of Figure 3.16b from Figure 3.17b gives the final result illustrated in Figure 3.17c. This triangular BMD is the same shape as that due to an applied point load at B, that is, the differential temperature distribution induces a downward reaction at that point.

EXAMPLE 3.8 DIFFERENTIAL TEMPERATURE CHANGE IN CONTINUOUS BEAM

The three-span beam illustrated in Figure 3.18 is subjected to an increase in temperature, which varies linearly from a maximum of 20° at the top to 10° at the bottom. The depth of the beam is h, and the centroid is at mid-depth; the elastic modulus is E, and the second moment of area is I. It is required to find the BMD due to these temperature increases, given that the coefficient of thermal expansion is α.

The temperature distribution is first converted into a strain distribution by multiplying by the coefficient of thermal expansion, α. The distribution is then resolved into two components, axial strain and bending strain, as illustrated in Figure 3.19. As the beam is free to expand, the

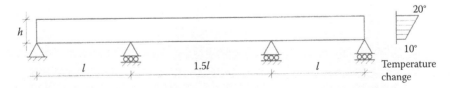

Figure 3.18 Differential temperature example.

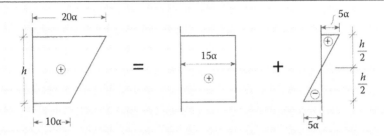

Figure 3.19 Resolution of applied change in strain into axial and bending components.

axial component will result in a free expansion, that is, a strain but no stress. The bending component will result in some moment but not as much as would occur if the beam were totally prevented from bending. The BMD will be determined using the method of equivalent loads.

Stage A – Calculate the equivalent loads and the associated stresses: In this example, the curvature is, from Figure 3.19,

$$\kappa = \frac{\varepsilon}{y} = \frac{5\alpha}{(h/2)} = \frac{10\alpha}{h}$$

Hence, the equivalent moment becomes

$$M = EI\kappa = EI\frac{10\alpha}{h}$$

Thus, the equivalent loads and associated BMD are as illustrated in Figure 3.20.

Stage B – Analyse for the effects of the equivalent loads: The frame is analysed for the loading of Figure 3.20a. The resulting BMD is illustrated in Figure 3.21a.

Stage C – Subtract the associated stresses: Subtracting the associated BMD of Figure 3.20b from Figure 3.21a gives the final result illustrated in Figure 3.21b. This is the BMD due to the differential temperature increase.

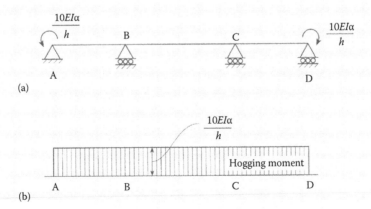

Figure 3.20 Application of equivalent loads method: (a) equivalent loads; (b) associated BMD.

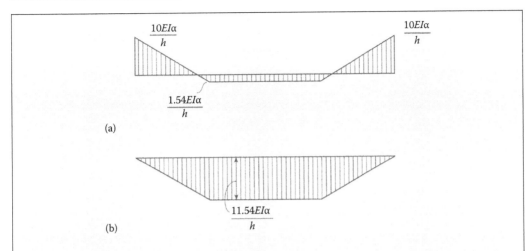

Figure 3.21 Completion of equivalent loads method: (a) BMD due to analysis for equivalent loading; (b) final BMD after subtraction of associated BMD.

EXAMPLE 3.9   BRIDGE DIAPHRAGM

The bridge diaphragm illustrated in Figures 3.22a and b is subjected to the differential increase in temperature shown in Figure 3.22c. It is required to determine if there will be uplift at B due

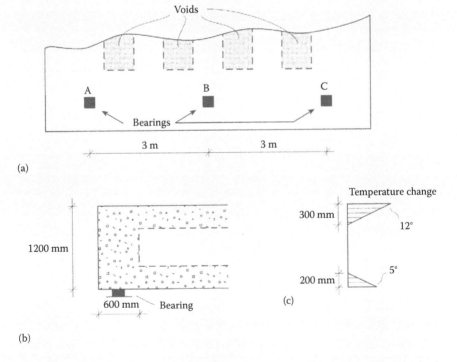

Figure 3.22 Bridge diaphragm example: (a) plan of geometry; (b) section through diaphragm; (c) applied temperature distribution.

to combined temperature and dead load. The upward reaction from the bearing due to the dead load is 300 kN, the coefficient of thermal expansion is 12×10^{-6} and the modulus of elasticity is 35,000 N/mm².

The cross section and temperature distribution for this example are identical to those of Example 2.2 (Chapter 2). Referring to that example, the equivalent loading is a force of 580 kN and a moment of 160 kNm, of which only the moment is of relevance here.

To determine the reaction due to this moment, the structure is analysed for the loading illustrated in Figure 3.23a. The associated BMD is illustrated in Figure 3.23b. A two-dimensional analysis, such as finite element or grillage, should now be used to find the reactions (see Chapters 5 and 6). However, a preliminary check is carried out here using a simple beam assumption. By symmetry, point B does not rotate and is effectively fixed. Hence (as in Example 3.7), the BMD due to applied loading is as illustrated in Figure 3.23c. Subtracting the associated BMD gives the

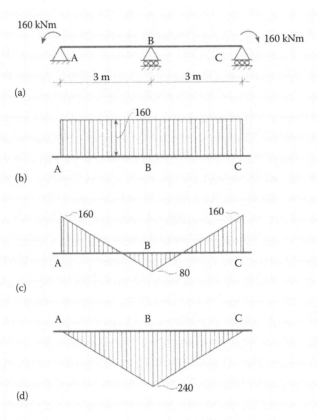

Figure 3.23 Analysis to determine effect of imposed differential temperature: (a) equivalent loading; (b) associated BMD; (c) results of analysis; (d) final BMD.

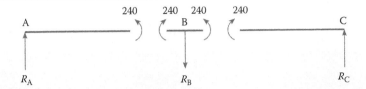

Figure 3.24 Free body diagram for diaphragm beam.

final BMD illustrated in Figure 3.23d. The reactions at A and C can be found from the free body diagram illustrated in Figure 3.24:

$$R_A = R_C = 240/3 = 80 \text{ kN}$$

Hence, the reaction at B is 80 + 80 = 160 kN. As the reaction due to dead load exceeds this value, there is no uplift of this bearing due to the differential temperature change.

EXAMPLE 3.10   DIFFERENTIAL TEMPERATURE IN BRIDGE OF NON-RECTANGULAR SECTION

The beam-and-slab bridge whose section and temperature loading is described in Example 2.4 consists of two 10-m spans. It is required to determine the maximum stresses due to the differential temperature change. In Example 2.4, it was established that the equivalent moment due to the temperature change is $-0.72\alpha E$ for half of the bridge. Using the method of equivalent loads, we have the following:

Stage A: The equivalent loads are illustrated in Figure 3.25a and the associated BMD in Figure 3.25b.

Stage B: Analysis for these loads gives the BMD illustrated in Figure 3.25c.

Stage C: Subtracting the associated BMD of Figure 3.25b from Figure 3.25c gives the final distribution of moment due to restrained bending illustrated in Figure 3.25d.

Thus, a sagging bending moment is induced over the central support, B, of $1.08\alpha E$, which gives stresses (tension positive) of $-5.29\alpha E$ and $11.35\alpha E$ at the top and bottom fibres, respectively. It was established in Example 2.4 that the residual stresses are $-5.37\alpha E$ and $-5.47\alpha E$ (restraint to expansion induces compression at the extreme fibres). Hence, the total stress at the top fibre is $-5.29\alpha E - 5.37\alpha E = -10.66\alpha E$. At the bottom fibre, the total stress is $11.35\alpha E - 5.47\alpha E = 5.88\alpha E$.

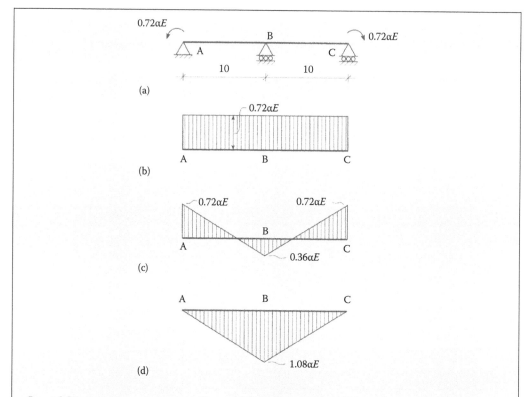

Figure 3.25 Analysis to determine effect of differential temperature change: (a) equivalent loading; (b) associated BMD; (c) results of analysis; (d) final BMD.

EXAMPLE 3.11   VARIABLE SECTION BRIDGE

Figure 3.26a shows the elevation of a pedestrian bridge, whereas Figures 3.26b, c and d show sections through it. The deck is subjected to the differential *decreases* in temperature shown in the figure. The bridge is first restrained when its temperature is somewhere between 5°C and 25°C, and the minimum temperature attained during its design life is −15°C. It is required to determine the equivalent loading and the associated stress distributions given a coefficient of thermal expansion, $\alpha = 12 \times 10^{-6}/°C$ and a modulus of elasticity, $E = 35 \times 10^{6}$ kN/m².

By summing moments of area about a point, it is found that the centroids are 0.5 and 1.033 m below the top fibre for the solid and hollow sections, respectively (Figures 3.26b–d). Summing products of stress and area in Figure 3.26b gives the equivalent force (positive tension) on the solid section due to the differential temperature distribution:

$$F_i = -\tfrac{1}{2}(2\alpha E)(2.6 \times 0.2) - \tfrac{1}{2}(8\alpha E)(2.6 \times 0.2)$$
$$= -1092 \text{ kN}$$

However, the Eurocode states that this uniform component of the differential temperature distribution is already included in the uniform temperature range specified and does not need to be considered here.

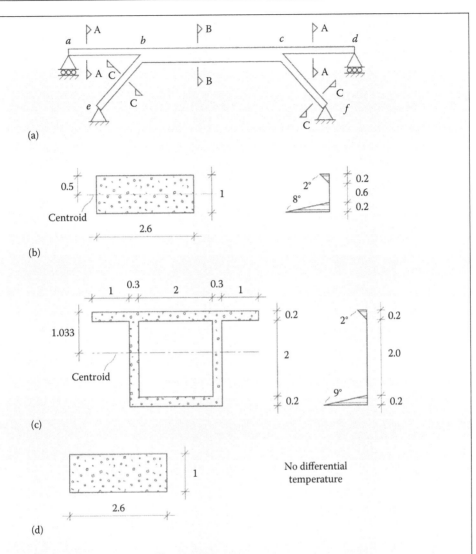

Figure 3.26 Pedestrian bridge: (a) elevation; (b) section A–A and corresponding imposed temperature distribution; (c) section B–B and corresponding imposed temperature distribution; (d) section C–C.

The corresponding equivalent moment (positive sag) is

$$M_1 = \tfrac{1}{2}(2\alpha E)(2.6 \times 0.2)\left(0.5 - \frac{0.2}{3}\right) - \tfrac{1}{2}(8\alpha E)(2.6 \times 0.2)\left(0.5 - \frac{0.2}{3}\right)$$

$$= -284\,\text{kNm}$$

In the hollow section, the equivalent force due to the differential temperature distribution is

$$F_2 = -\tfrac{1}{2}(2\alpha E)(4.6 \times 0.2) - \tfrac{1}{2}(9\alpha E)(2.6 \times 0.2)$$

$$= -1369\,\text{kN}$$

(but does not need to be considered) and the equivalent moment is

$$M_2 = \tfrac{1}{2}(2\alpha E)(4.6 \times 0.2)\left(1.033 - \frac{0.2}{3}\right) - \tfrac{1}{2}(9\alpha E)(2.6 \times 0.2)\left(2.4 - 1.033 - \frac{0.2}{3}\right)$$

$$= -905 \text{ kNm}$$

The maximum axial decrease in temperature is $(25-(-15)) = 40°C$, and the corresponding stress is $40\alpha E$. For the solid section of Figures 3.26b and d, the area is $2.6 \times 1 = 2.6 \text{ m}^2$, giving an equivalent force of

$$F_3 = -40\alpha E(2.6) = -43,680 \text{ kN}$$

For the hollow section, the area is 2.64 m^2 and the equivalent force is

$$F_4 = -40\alpha E(2.64) = -44,352 \text{ kN}$$

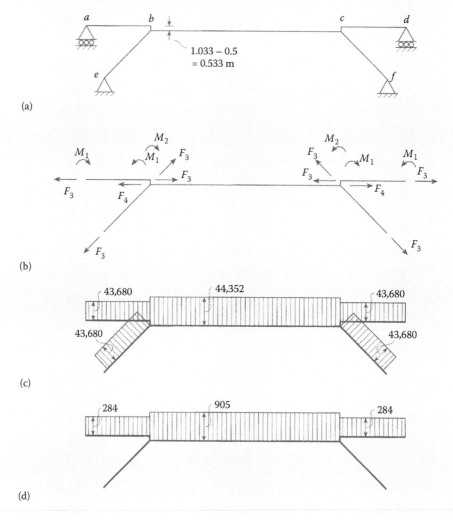

Figure 3.27 Model of pedestrian bridge: (a) geometry showing differences in level of centroids; (b) equivalent loading; (c) associated axial-force diagram; (d) associated BMD.

A model that allows for the difference in the level of the centroids is illustrated in Figure 3.27a. Note that the short vertical members at *b* and *c* could be assumed to have effectively infinite stiffness. However, using members with very large stiffnesses can generate numerical instability in a computer model. Therefore, a second moment of area several times as large as the maximum used elsewhere in the model (e.g. 10 times) generally provides sufficient accuracy without causing such problems. Noting that the axial effects apply to all members while the differential temperature distributions only apply to the deck (*abcd*), the equivalent loads are illustrated in Figure 3.27b. The associated axial force and BMDs are illustrated in Figures 3.27c and d, respectively. The bending moment and axial force distributions due to the temperature decreases can be found by analysing for the equivalent loading illustrated in Figure 3.27b and subtracting the associated distributions of Figures 3.27c and d from the results.

3.5.1 Temperature effects in three dimensions

When the temperature of a particle of material in a bridge is increased, the particle tends to expand in all three directions. Similarly, when a differential distribution of temperature is applied through the depth of a bridge slab, it tends to bend about both axes. If there is restraint to either or both rotations, bending moment results about both axes as will be illustrated in the following example.

EXAMPLE 3.12   DIFFERENTIAL TEMPERATURE

The slab bridge of Figure 3.28 is articulated as shown in Figure 3.28a to allow axial expansion in both the X and Y directions. However, for rotation, the bridge is two-span longitudinally and is therefore not able to bend freely. Further, there are three bearings transversely at the ends so that it is not able to bend freely transversely either.

The deck and cantilevers are subjected to the differential temperature increases illustrated in Figures 3.28c and d, respectively. It is required to determine the equivalent loading and the associated BMD due to this temperature change. The coefficient of thermal expansion is $9 \times 10^{-6}/°C$, and the modulus of elasticity is 32×10^6 kN/m². The specified temperature distributions are different in the cantilevers and the main deck of this bridge. However, for longitudinal bending, the bridge will tend to act as one unit, and bending will take place about the centroid. The location of this centroid is

$$\bar{x} = \frac{\{(10 \times 0.8) \times 0.4 + (2 \times 1.5 \times 0.25) \times 0.125\}}{\{(10 \times 0.8) + (2 \times 1.5 \times 0.25)\}} = 0.376 \text{ m}$$

below the top surface. The bridge deck is divided into parts, as illustrated in Figure 3.29, corresponding to the different parts of the temperature distribution, and the temperature changes are converted into stresses.

Taking moments about the centroid gives a longitudinal bending moment per metre on the main deck of

$$M_1 = -\frac{1}{2}(4320)(1 \times 0.35)\left(0.376 - \frac{0.35}{3}\right) + \frac{1}{2}(1440)(1 \times 0.2)\left(0.8 - 0.376 - \frac{0.2}{3}\right)$$

$$= -145 \text{ kNm/m}$$

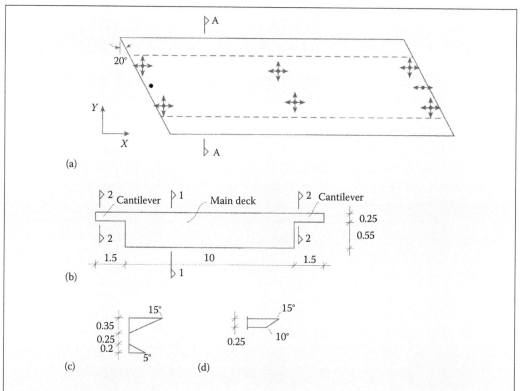

(a)

(b)

(c)   (d)

Figure 3.28 Slab bridge of Example 3.12: (a) plan showing directions of allowable movement at bearings; (b) section A–A; (c) imposed temperature distribution in deck (section 1–1); (d) imposed temperature distribution in cantilever (section 2–2).

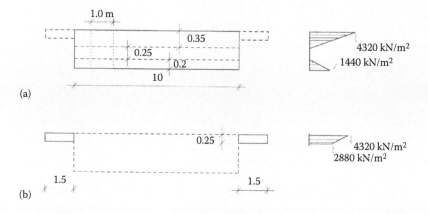

(a)

(b)

Figure 3.29 Cross section with associated distribution of imposed stress: (a) deck; (b) cantilevers.

The corresponding bending moment per metre on the cantilever is

$$M_2 = -(2880)(1 \times 0.25)\left(0.376 - \frac{0.25}{2}\right) - \tfrac{1}{2}(4320 - 2880)(1 \times 0.25)\left(0.376 - \frac{0.25}{3}\right)$$

$$= -233 \text{ kNm/m}$$

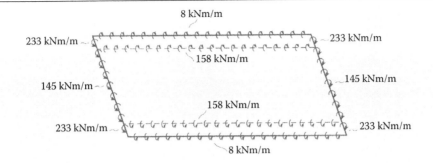

Figure 3.30 Equivalent loading due to temperature.

These equivalent longitudinal moments are illustrated in Figure 3.30.

The transverse direction is different from the longitudinal in that the cross section is rectangular everywhere. In the cantilever region, bending is about the centroid of the cantilever. The applied stress distribution is resolved into axial and bending components, as illustrated in Figure 3.31. The axial expansion is unrestrained while the bending stress distribution generates a moment per metre of

$$M_3 = -2\left\{ \tfrac{1}{2}(720)(1 \times 0.125)\left(0.125 - \frac{2}{3} \right) \right\}$$

$$= -8 \text{ kNm/m}$$

In the main deck, the differential distribution is applied to a 0.8 m deep rectangular section, giving a moment about the centroid of

$$M_4 = -\tfrac{1}{2}(4320)(1 \times 0.35)\left(0.4 - \frac{0.35}{3} \right) + \tfrac{1}{2}(1440)(1 \times 0.2)\left(0.4 - \frac{0.2}{3} \right)$$

$$= -166 \text{ kNm/m}$$

As M_3 is applied to the outside of the cantilever, only $(M_4 - M_3)$ needs to be applied at the deck/cantilever interface, as illustrated in Figure 3.30. As these applied moments generate distributions of longitudinal and transverse moment, there are two associated BMDs, as illustrated in Figure 3.32. As for the previous example, the problem is completed by analysing the slab and subtracting the associated BMDs from the solution.

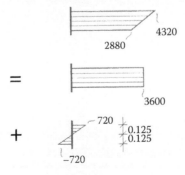

Figure 3.31 Resolution of imposed stress in cantilever into axial and bending components.

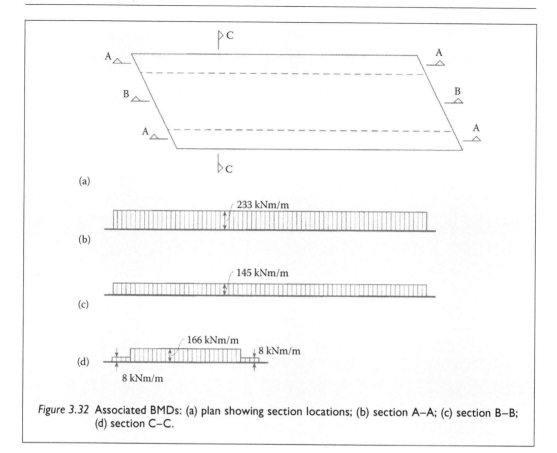

Figure 3.32 Associated BMDs: (a) plan showing section locations; (b) section A–A; (c) section B–B; (d) section C–C.

3.6 PRESTRESS

The effects of prestress in bridges are similar to the effects of temperature, and the same analysis techniques can be used for both. However, there is one important distinction. An unrestrained change in temperature results in a change in strain only and no change in stress. Prestress, on the other hand, results in changes of both stress and strain. For example, if a beam rests on a sliding bearing at one end, it can undergo axial changes in temperature without incurring any axial stress. However, prestressing that beam does (as is the objective) induce a distribution of stress. When the movements due to prestressing are unrestrained, the stress distributions are easily calculated, and analysis is not generally required. However, there are many bridge forms where the effects of prestress are restrained to some degree or other and where analysis is necessary.

EXAMPLE 3.13   FRAME SUBJECT TO AXIAL PRESTRESS

The frame of Figure 3.11, reproduced here as Figure 3.33, is subjected to a prestressing force along the centroid of the deck, ABC, of magnitude P. It is required to determine the net pre-stress force in the deck and the resulting BMD. The area of the deck is $500I_0/I^2$.

The frame is analysed for the two applied forces of magnitude P. Unlike temperature loading, there is no correction for associated stresses. The results of the analysis are given in Figure 3.34.

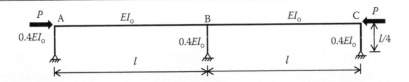

Figure 3.33 Frame subjected to prestress force.

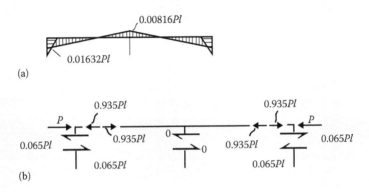

Figure 3.34 Effect of prestress force: (a) BMD; (b) internal shear and axial forces.

This example serves to illustrate the loss of prestress force that occurs in a frame due to the restraint offered by the piers. In this example, 6.5% of the applied force is lost as shear force in the piers. This is in addition to other sources of prestress loss such as that due to elastic shortening and friction.

It is also of importance to note the bending moment that is inadvertently induced by the prestress. Interestingly, this bending moment is independent of the elastic modulus and is therefore unaffected by creep.

EXAMPLE 3.14   ANALYSIS FOR ECCENTRIC PRESTRESSING

The beam illustrated in Figure 3.35 is prestressed with a straight tendon at an eccentricity, e, from the centroid with a prestress force, P. (It should be noted that this is an unrealistic prestressing arrangement.) It is required to determine the induced distributions of axial force and bending moment.

The method of equivalent loads is applicable to prestress just as it is to temperature. The only difference is that, as prestress generates stress as well as strain, it is not appropriate to deduct

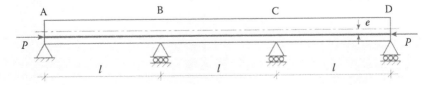

Figure 3.35 Beam subjected to eccentric prestress force.

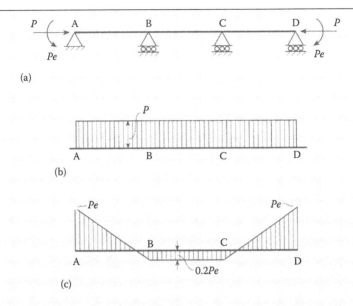

Figure 3.36 Prestressed beam of Example 3.14: (a) equivalent loads; (b) axial-force diagram due to prestress; (c) BMD due to prestress.

the associated stresses from the analysis results as was necessary in temperature analysis. In this example, the prestress force is applied at an eccentricity to the centroid. This is equivalent to applying a moment alongside the force as illustrated in Figure 3.36a. The axial force diagram is clearly as illustrated in Figure 3.36b. To determine the BMD, however, is not so straightforward as the beam is not free to lift off the supports at B and C. An analysis for the equivalent loads of Figure 3.36a gives the BMD of Figure 3.36c.

It is interesting to note from this example that the effect of the tendon below the centroid is to generate sagging moment in the central span. In a simply supported beam, a tendon below the centroid generates hogging moment.

EXAMPLE 3.15   PROFILED TENDONS

In most post-tensioned bridges, the tendons are profiled using a combination of straight portions and parabolic curves. For preliminary design purposes, the actual profiles are sometimes approximated by ignoring the transition curves over the internal supports, as illustrated in Figure 3.37. For this beam, it is required to find the BMD due to a prestress force, P. A parabolic profile generates a uniform loading, the intensity of which can be determined by considering equilibrium of forces at the ends of the parabola. (This was covered in greater detail in Chapter 2.) For the parabola in span AB, the slope is found by differentiating the equation as follows:

$$y = \frac{0.1x^2}{l} - 0.08x$$

$$\Rightarrow \quad \text{slope} = \frac{dy}{dx} = \frac{0.2x}{l} - 0.08$$

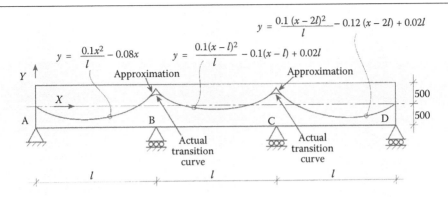

Figure 3.37 Beam with profiled prestressing tendon.

At A, $x = 0$ and the slope becomes -0.08. From Figure 3.38a, it can be seen that the vertical component of the prestressing force at A is $P\sin\theta_1 \approx P\tan\theta_1 = 0.08P$. Similarly, at $x = l$, the slope is 0.12, and the vertical component of prestress is $0.12P$. Hence, the equilibrium of vertical forces requires a uniform loading of intensity:

$$w_{AB} = (0.08P + 0.12P)/l = 0.2P/l$$

At the ends of BC, the vertical components of prestress force can be found similarly. They are both equal to $0.1P$, and the intensity of loading is, coincidentally, $w_{BC} = 0.2P/l$. In CD, the intensity is, by symmetry, $w_{CD} = w_{AB} = 0.2P/l$. Thus, the complete equivalent loading due to prestress is as illustrated in Figure 3.38b. The beam is analysed for this loading and the results presented in Figure 3.39. This example serves to illustrate that the effect of profiled prestressing tendons

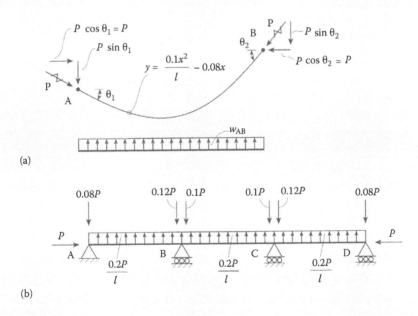

Figure 3.38 Equivalent loading due to profiled tendon: (a) equivalent forces in span AB; (b) summary of all equivalent forces on beam.

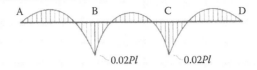

Figure 3.39 BMD due to prestress.

can be quite similar to the effect of self-weight in that it applies a uniform loading throughout the beam. The obvious difference is that typical prestress loading is in the opposite direction to loading due to self-weight.

Equivalent loading due to prestress can become complex when there are several sources of equivalent loads such as those due to prestress losses and variable beam geometry. In such cases, an alternative and perhaps simpler approach is the method of incremental moments. This consists simply of discretizing the bridge into a number of segments and calculating applied moments = *force × eccentricity* at each point. The increments of moment between each of these points are then calculated and applied as equivalent loads to the structure. The method is illustrated using the following example.

EXAMPLE 3.16   METHOD OF INCREMENTAL MOMENT

Use the method of incremental moments to find the moment due to prestress for the two-span beam of Figure 3.40, where $e = 0.3$ and $s = 0.5$ m. The tendon is post-tensioned simultaneously from A and C, so, due to friction losses, the prestress force varies linearly from P at A and C to $0.9P$ at B.

From Equation 2.14, the tendon profile in the first span is given by

$$y = e_A + (e_B - e_A)\frac{x}{l} - 4s\left[\frac{x}{l} - \left(\frac{x}{l}\right)^2\right]$$

where x is the distance from the start of the span, and y is the distance from the beam centroid. Taking the first span as typical, $e_A = 0$, $e_B = e = 0.3$ and $s = 0.5$, giving

$$y = e\frac{x}{l} - 4s\left[\frac{x}{l} - \left(\frac{x}{l}\right)^2\right]$$

$$= -1.7\frac{x}{l} + 2\left(\frac{x}{l}\right)^2$$

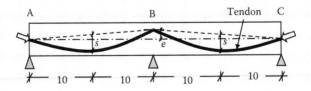

Figure 3.40 Tendon profile.

The beam is discretized at 1/8 points and the coordinates of the tendon evaluated using this equation. The results are presented in the second column of Table 3.1. As the y-coordinates are measured from the centroid, they are the eccentricities. The linearly varying prestress forces are given in the third column of Table 3.1. The applied prestress moments are simply the products of the forces and eccentricities at each point; these are given in the fourth column of Table 3.1. Finally, the incremental moments are found by subtracting successive applied moments and are presented in the fifth column.

The applied moments for the first 10 m of the beam are illustrated in Figure 3.41a and the incremental moments. The latter are applied to the structure as shown in Figure 3.41b. The resulting BMD is given in Figure 3.41c. It can be seen that the final BMD due to prestress is

Table 3.1 Calculation of incremental moments

x/l	y = ecc	Prestress force	Applied prestress moment	Incremental moments
0	0	0	0	
$\frac{1}{16}$				−0.179P
$\frac{1}{8}$	−0.181	0.988P	−0.179P	
$\frac{3}{16}$				−0.114P
$\frac{1}{4}$	−0.300	0.975P	−0.293P	
$\frac{5}{16}$				−0.050P
$\frac{3}{8}$	−0.356	0.963P	−0.343P	
$\frac{7}{16}$				0.010P
$\frac{1}{2}$	−0.350	0.95P	−0.333P	
$\frac{9}{16}$				0.069P
$\frac{5}{8}$	−0.281	0.938P	−0.264P	
$\frac{11}{16}$				0.125P
$\frac{3}{4}$	−0.150	0.925P	−0.139P	
$\frac{13}{16}$				0.179P
$\frac{7}{8}$	0.044	0.913P	0.040P	
$\frac{15}{16}$				0.230P
1	0.300	0.9P	0.270P	

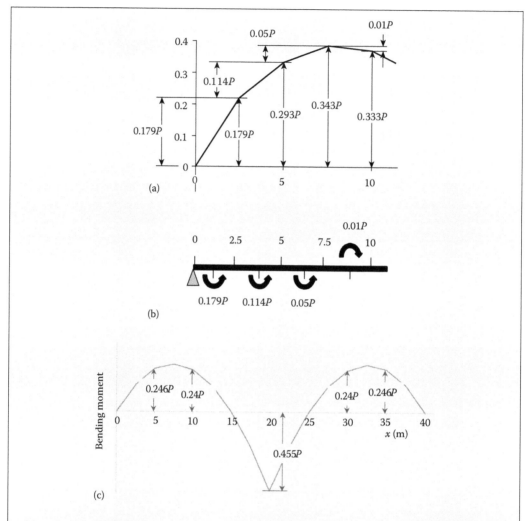

Figure 3.41 Method of incremental moments: (a) segment of applied BMD and incremental moments (differences); (b) segment of equivalent loading due to prestress; (c) BMD due to prestress.

significantly different from the applied moments. For example, at the central support, B, the product of prestress force and eccentricity is 0.27P, but the final moment at that point is 0.455P. This difference between applied and actual occurs in indeterminate structures and is known as the parasitic or secondary moment.

3.7 ANALYSIS FOR THE EFFECTS OF CREEP

Creep in concrete means that strain continues to increase when stress is kept constant over an extended period. In a determinate structure constructed at one instant in time, this is not a problem. For example, an in situ reinforced concrete, simply supported bridge will creep over time, but this does not cause any change in the stresses. At each cross section, the bending moment due to permanent loading will cause a triangular distribution of stresses through the depth. There will be a corresponding triangular distribution of strain and an associated curvature. Creep will cause the strains and the associated curvature to increase over time. This will cause the deflections to increase. but, as the movements are not constrained, it will not generate any stress or bending moment.

In determinate structures constructed in multiple stages, creep can have an effect on the stresses. For example, simply supported bridges are sometimes made up of precast inverted-T-beams and in situ concrete (see Figure 1.3). When the in situ concrete is first poured, it has no strength, so the entire self-weight is carried by the precast beams. However, over time, the bridge creeps and some of the stresses due to self-weight come to be carried by the composite structure – precast and in situ.

Creep has a more profound influence on stresses when the structural form of an indeterminate bridge changes during construction. This happens, for example, in the case of partially continuous beam/slabs (Section 1.4.4 and Figure 1.19) and in balanced cantilever construction (Section 1.4.6 and Figure 1.25).

The total creep strain, in a situation of constant stress, σ_{el}, is given by $\varphi\sigma_{el}/E$, where φ is the creep coefficient, and E is the elastic modulus. The rate of creep method of analysis, credited to Dischinger, allows this type of calculation to be applied incrementally. It is generally described as being less accurate than other approaches but is widely used when the influence of creep is small. In a given short interval of time, the rate of creep method gives the increment of creep strain as the sum of an incremental creep part and a relieving part due to any reduction in the stress:

$$d\varepsilon = \sigma\frac{d\varphi}{E} - \frac{d\sigma}{E} \tag{3.1}$$

where σ is the current stress, $d\sigma$ is the change in stress in this step, E is Young's modulus and $d\varphi$ is the creep coefficient. This approach is first illustrated with a simple theoretical example.

EXAMPLE 3.17   LOAD SHARING DUE TO CREEP

The structure of Figure 3.42a consists initially of just spring A with stiffness k. A force, P, is applied and results in an internal axial force of P in that spring. Spring B is then added, also with stiffness k (Figure 3.42b). Both springs are susceptible to creep, which will have the effect of transferring axial force from spring A to spring B over time. The problem can be addressed by considering many small increments of strain loading.

The first stage of loading is elastic and trivial – the applied force results in an internal axial force of P in spring A. Subsequently, there are, say, n stages of creep loading, noting that each strain step will correspond to a different length of time. For any stage, the total creep strain

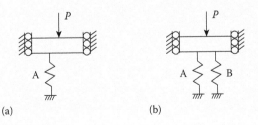

(a)   (b)

Figure 3.42 Load sharing between springs due to creep: (a) initial condition; (b) after insertion of second spring.

is proportional to the force. In stage 2, this is P in spring A and zero in spring B. The creep strain in spring A is proportional to $\varphi P/n = \psi P$, say. The method of equivalent loads is adopted as was used in analysis for the effects of temperature changes. Hence, in stage 2, a force of ψP is applied to spring A. As the springs have equal stiffness, the load is carried equally by each of them – the internal forces resulting from this 'analysis' are each $\frac{1}{2}\psi P$. This is summarised in Table 3.2. The axial force in spring A at the end of stage 2 is the starting force, P, the results of the analysis, $\frac{1}{2}\psi P$, minus the associated axial force, ψP. Hence, the force in that spring at the start of stage 3 is, $P + \frac{1}{2}\psi P - \psi P = (1 - \frac{1}{2}\psi)P$. Spring B has no axial force at the start of stage 2 and gains $\frac{1}{2}\psi P$ from the analysis.

At the start of Stage 3, the axial force in spring A is $(1 - \frac{1}{2}\psi)P$, so the equivalent load is ψ times that. For this stage, spring B also has a small force, so it too is subject to creep and has an equivalent load of $\psi \times \frac{1}{2}\psi P = \frac{1}{2}\psi^2 P$. Hence, the total equivalent load is $(\psi - \frac{1}{2}\psi^2)P + \frac{1}{2}\psi^2 P = \psi P$ (as before), and the results of the 'analysis' is an internal force of $\frac{1}{2}\psi P$ in each spring. At the end of this stage, the force in spring A is $(1 - \frac{1}{2}\psi)P + \frac{1}{2}\psi P - (\psi - \frac{1}{2}\psi^2)P = (1 - \psi + \frac{1}{2}\psi^2)P$. The force in spring B is $\frac{1}{2}\psi P + \frac{1}{2}\psi P - \frac{1}{2}\psi^2 P = (\psi - \frac{1}{2}\psi^2)P$. This analysis assumes that the rate of creep, φ and ψ, are constant at each stage, which may not always be the case. Figure 3.43 illustrates the progression of the forces in each spring through time, assuming equal creep strain stages in geometrically increasing time intervals.

Table 3.2 Summary of analysis for creep in two-spring problem of Figure 3.42

	Spring A			Spring B		
	Axial force at start of stage	Equivalent load = associated axial force	Results of analysis	Axial force at start of stage	Equivalent load = associated axial force	Results of analysis
Stage 1	P	–	P	–	–	–
Stage 2	P	ψP	$\frac{1}{2}\psi P$	0	0	$\frac{1}{2}\psi P$
Stage 3	$(1 - \frac{1}{2}\psi)P$	$(\psi - \frac{1}{2}\psi^2)P$	$\frac{1}{2}\psi P$	$\frac{1}{2}\psi P$	$\frac{1}{2}\psi^2 P$	$\frac{1}{2}\psi P$
Stage 4	$(1 - \psi + \frac{1}{2}\psi^2)P$	•	•	$(\psi - \frac{1}{2}\psi^2)P$	•	•
•	•	•	•	•	•	•
•	•	•	•	•	•	•

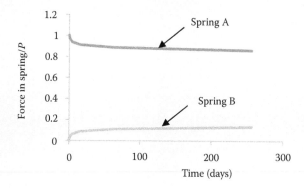

Figure 3.43 Progression of spring forces with time in days.

This simple example illustrates the way in which creep can affect structures that are constructed in stages. It can result in load sharing, as shown, with stress migrating from one part of a structure to another. Creep affects bending structures just as it affects structures consisting of axially loaded members. In bending structures, the stress is still axial, and bending moment describes the distribution of axial stress at a particular section on the structure. It follows that creep causes migrations of moment from one part of a structure to another. The phenomenon can be quantified by applying a number of small increments of creep strain. The method of equivalent loads can be used for this purpose, as was used for the analysis of bridges subject to differential temperature changes. This is illustrated in Example 3.18.

EXAMPLE 3.18 EFFECT OF CREEP ON BENDING MOMENT IN A PARTIALLY CONTINUOUS BEAM

A structure consists initially of two precast concrete beams, as illustrated in Figure 3.44a, and is supported on temporary bearings at B. In situ concrete is poured to tie the beams together, including a full depth diaphragm at B – Figure 3.44b. The temporary bearings are then removed and replaced with a permanent one. It is required to find the BMD due to self-weight of 20 kN/m, assuming a rate of creep for both precast and in situ concrete of $\varphi = 0.8$ (i.e. total creep strain equal to 80% of elastic strain).

When the structure consists of simply supported beams, its BMD is of the shape illustrated in Figure 3.45a; each beam carries its self-weight, which results in a parabolic BMD. When the in situ concrete is first poured, it has no structural strength, so its weight is also carried by the precast beams. Hence, all the self-weight of the bridge – both of precast and in situ

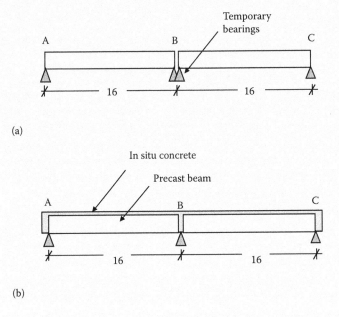

(a)

(b)

Figure 3.44 Two-span partially continuous beam: (a) initial structure consists of two simply supported precast beams; (b) final structure consists of precast beams and in situ concrete and is continuous over two spans.

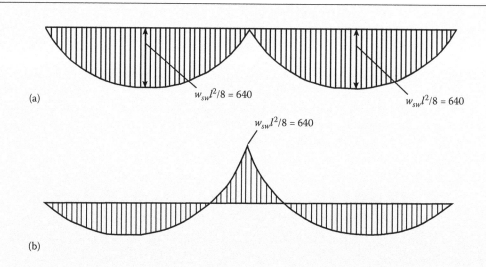

$w_{sw}l^2/8 = 640$

$w_{sw}l^2/8 = 640$

$w_{sw}l^2/8 = 640$

(a)

(b)

Figure 3.45 BMDs due to self-weight: (a) actual initial BMD; (b) BMD that would exist if the structure were continuous from the beginning.

concrete – generates the BMD of Figure 3.45a. If the structure were continuous when the self-weight was applied, then the BMD would be that of Figure 3.45b. Creep has the effect of changing the initial BMD from that of Figure 3.45a into a shape somewhere between that of Figures 3.45a and b.

The analysis will be carried out using one elastic stage and n stages of incremental creep loading. After the elastic stage of loading, the BMD is as illustrated in Figure 3.45a. In the second stage of loading, the creep strain to be applied is proportional to $\varphi\sigma/n$, where σ is the sustained stress. As bending moment is a distribution of stress, the curvature due to creep is proportional to the sustained bending moment. Hence, the equivalent loading due to creep is $M_{eq} = \varphi M_{sw}/n$, where M_{sw} is the moment due to self-weight.

Discretizing each beam into eight segments, the moments in the first beam due to self-weight are given in the second column of Table 3.3. Taking $n = 10$ stages of creep, and recalling that $\varphi = 0.8$, the equivalent moments due to creep in the second stage are $M_{eq} = 0.08M_{sw}$. These are given in the third column of the table. The analysis will be done using the method of equivalent loading, that is, the incremental moments given in the fourth column are applied to the structure, it is analysed to find the BMD, and the associated BMD is subtracted from the result. (The process of applying the incremental moments is similar to that described in Example 3.16.) The results of the analysis are given in the final column of the table.

Stage 3 of the analysis starts with a recalculation of the sustained moments. The sustained moments are now the previous values, M_{sw}, plus the results of the stage 2 analysis (last column of Table 3.3), minus the associated moments, which are equal to the applied equivalent moments, M_{eq} (third column of Table 3.3). These updated sustained moments are given in the second column of Table 3.4 where it can be seen that there is a significant change, with a hogging moment starting to emerge at the right end of the first beam (point B). The applied moments in this stage are 0.08 times these sustained moments, as shown in the table. The process is repeated for 10 stages of creep (11 stages in total). The updated BMD at the end of each stage is shown in Figure 3.46. The final moment at B is 356 kN m.

Table 3.3 Stage 2 analysis for creep in partially continuous beam (1st creep stage)

x (m)	Moment due to self-weight, M_{sw}	Applied moments in stage 2, M_{eq}	Incremental moments	Analysis results
0	0	0		0
1			22.4	
2	280	22.4		16.10
3			16.0	
4	480	38.4		25.80
5			9.6	
6	600	48.0		29.10
7			3.2	
8	640	51.2		26.00
9			−3.2	
10	600	48.0		16.50
11			−9.6	
12	480	38.4		0.60
13			−16	
14	280	22.4		−21.70
15			−22.4	
16	0	0		−50.40

Table 3.4 Stage 3 analysis for creep in partially continuous beam

x (m)	Sustained moment in stage 3	Applied moments in stage 3	Incremental moments	Analysis results
0	0	0		0
1			21.9	
2	274	21.9		16.10
3			15.5	
4	467	37.4		25.80
5			9.1	
6	581	46.5		29.10
7			2.7	
8	615	49.2		26.00
9			−3.7	
10	569	45.5		16.50
11			−10.1	
12	442	35.4		0.61
13			−16.5	
14	236	18.9		−21.69
15			−22.9	
16	−50	0		−50.39

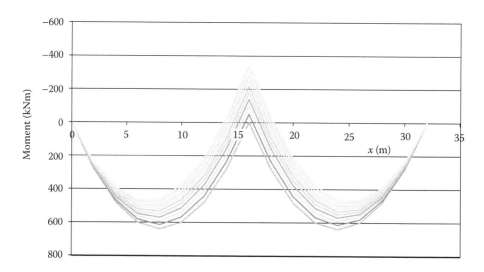

Figure 3.46 BMDs at different stages in the life of the two-span, partially continuous beam of Figure 3.44.

Hewson (2003) gives a simpler formula for bending moment in bridges of this form:

$$M = M_{ini} + (M_{fin} - M_{ini})(1 - e^{-\varphi})$$

where M_{ini} is the initial BMD, and M_{fin} is the BMD that would have resulted if the loading were applied to the final geometry. Applying this formula to the central support moment in Example 3.18 gives

$$M = 0 + (640 - 0)(1 - e^{-0.8}) = 352 \text{ kNm.}$$

Chapter 4

Integral bridges

4.1 INTRODUCTION

Integral bridges are those where the superstructure and substructures are continuous or integral with each other. While the concept is well established, many bridges built in the 1960s and 1970s were articulated with expansion joints and bearings to separate the superstructure from the substructure and the surrounding soil. In the 1980s and 1990s, many of these articulated bridges required rehabilitation due to serviceability problems associated with the joints. As a result, integral construction has become popular in recent decades, and this trend seems likely to continue into the future. It is particularly popular for shorter bridges and those without too much skew. In the United Kingdom, BD57/01 specifies that integral construction be generally used for road bridges with lengths up to 60 m and with up to 30° skew.

4.1.1 Integral construction

There are many variations on the basic integral bridge. In the bridge of Figure 4.1a, the deck is composed of separate precast beams in each span. While in the past, such a deck might have had a joint over the central support, in the more durable form of construction illustrated, it is made continuous over the support using in situ concrete. A bridge is shown in Figure 4.1b in which the deck is continuous over the internal support and integral with the abutments at the ends. Figure 4.1c illustrates another variation; this bridge is integral with both the abutments and the intermediate pier.

While there are considerable durability advantages in removing joints and bearings, their removal does affect the bridge behaviour. Specifically, expansion and contraction of the deck are restrained with the result that additional stresses are induced, which must be resisted by the bridge structure. The most obvious cause of expansion or contraction in bridges of all forms is temperature change, but other causes exist such as shrinkage in concrete bridges. In prestress concrete decks, elastic shortening and creep also occur. A simple integral bridge is illustrated in Figure 4.2a. If the bases of the abutments are not free to slide, deck contraction induces the deformed shape illustrated in Figure 4.2b and the bending moment diagram of Figure 4.2c. Partial sliding restraint at the bases of the abutments results in the deformed shape of Figure 4.2d and a bending moment diagram, which is similar in shape to Figure 4.2c but of a lesser magnitude.

Time-dependent contractions in concrete bridge decks induce bending moments in integral bridges. While the magnitude of creep contraction is time-dependent, creep also has the effect of relieving the induced bending moments over time. The net effect of this is that moments induced by creep contraction are often small. Shrinkage strain increases with time, but the resulting moments are also reduced by creep.

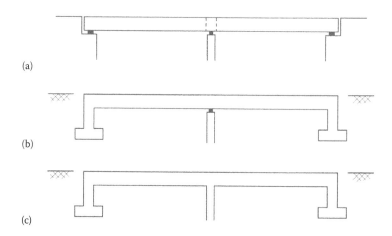

Figure 4.1 Integral bridges: (a) precast beams made integral over the interior support; (b) deck continuous over interior support and integral with abutments; (c) deck integral with abutments and pier.

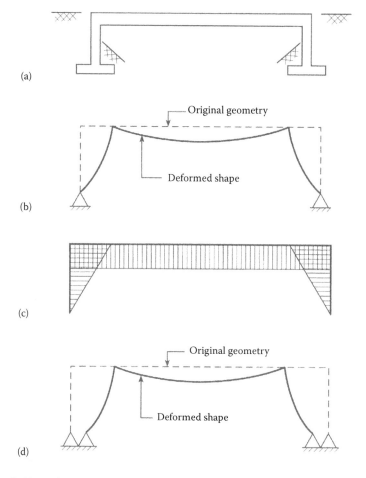

Figure 4.2 Frame bridge subject to contraction: (a) geometry; (b) deformed shape if bases are restrained against sliding; (c) bending moment diagram; (d) deformed shape if bases are partially restrained against sliding.

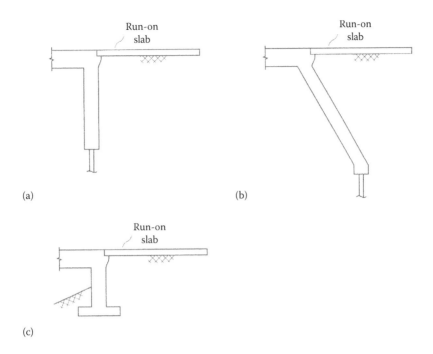

Figure 4.3 Ends of integral bridges: (a) deep vertical abutment; (b) deep inclined abutment; (c) bank seat abutment.

Elastic shortening occurs in post-tensioned, prestressed concrete decks during the application of prestress. If the deck is integral with the supports at the time of stressing, bending moments are induced. On the other hand, many integral bridges are constructed from precast, pre-tensioned beams, and the bridge is not made integral until after the pre-tensioning process is complete. In such cases, no bending moments are induced by the elastic shortening.

Temperature changes are another major source of deck expansion and contraction. Temperature can be viewed as having a seasonal and hence long-term component as well as a daily or short-term component. The resistance of an integral bridge to movement of any type depends largely on the form of construction of the substructures. Three alternative forms are illustrated in Figure 4.3. In each case, a run-on slab is shown behind the abutment. These are commonly placed over the transition zone between the bridge and the adjacent soil, which generally consists of granular backfill material. Figure 4.3a and b shows two bridges that are integral with high supporting abutments and piled foundations. In such a case, a reduction in lateral restraint can be achieved by using driven steel H-section piles with their axes orientated so that the stiffness of the weaker axis resists the movement. An alternative form of integral construction is one in which abutments sit on strip foundations like the small bank seat abutment illustrated in Figure 4.3c. Minimising the sliding resistance at the base of these foundations helps to reduce the lateral restraint. Care should be taken in the design to ensure that bank seats have sufficient weight in order to avoid uplift from applied loads in other spans.

4.1.2 Lateral earth pressures on abutments

The lateral earth pressures (σ_h) that the abutments of integral bridges should be designed for are those that take place during the maximum expansion of the bridge deck. The expansion has the effect of pushing the abutment laterally into the backfill. The resulting earth pressures developed on the abutment are dependent on the stiffness and strength of the backfill and on the amount of movement of the abutment.

The maximum lateral earth pressure that can be sustained by the backfill is termed the passive pressure (σ_{hp}), which, for dry backfill at a depth z and no surcharge at ground level, is given by the expression

$$\sigma_{hp}(z) = K_p\,\gamma_{soil}\,z \tag{4.1}$$

where K_p is the coefficient of passive pressure, and γ_{soil} is the unit weight of the backfill. The coefficient K_p may be estimated from Figure 4.4 for a given angle of internal friction of the backfill ϕ' and a given δ_a/ϕ' ratio, where δ_a is the angle of interface friction between the abutment and backfill. One design approach would be to use Equation 4.1 directly to determine the maximum lateral pressure distribution on the abutment. This approach, however, is generally overly conservative as abutment movements are usually significantly less than those required to generate passive pressures.

The preferred approach is one involving an appropriate soil/structure interaction analysis, which takes due account of the stiffness of the soil. Such an approach is described later in this chapter. A third (and commonly used) approach relates the pressure distribution on the abutment to the degree of mobilisation of its maximum (or passive) lateral capacity. This method is proposed in the guidelines of the UK Highway Agency's BA42/96 (2003), which assumes that the lateral stress at any depth z, $\sigma_h(z)$, is proportional to the vertical stress ($\gamma_{soil}\,z$):

$$\sigma_h(z) = K^*\gamma_{soil}\,z \tag{4.2}$$

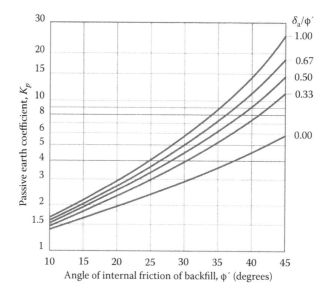

Figure 4.4 Coefficients of passive earth pressure (horizontal component) for horizontally retained surface. (From Caquot, A. and Kersiel, J., *Tables for the Calculation of Passive Pressure, Active Pressure and Bearing Capacity of Foundations* (translated from French by M.A. Bec), Gauthier-Villars, Paris, 1948.)

The constant of proportionality (K^*) is a function of the displacement at the top of the abutment (d) and the type of abutment. BA42/96 (2003) gives the following expressions for K^* (where K_0 is the coefficient of earth pressure at rest):

$$K^* = \text{Min}\left[K_p, \ K_0 + \left(\frac{d}{0.025H} \right)^{0.4} K_p \right] \qquad \text{Bank seats} \qquad (4.3)$$

$$K^* = \text{Min}\left[K_p, \left(\frac{d}{0.05H} \right)^{0.4} K_p \right] \quad \text{for } z \leq \xi H \quad \text{Full height abutments} \qquad (4.4)$$

The value of ξ is 0.5 for a full-height frame abutment and 0.66 for a full-height embedded wall abutment. At depths greater than ξH, $\sigma_h(z)$ is taken as the maximum of the values of $\sigma_h(z)$ evaluated using Equation 4.4 at $z = \xi H$ and the (normally consolidated) at-rest lateral stress given as

$$\sigma_{h0}(z) = (1 - \sin\phi')\gamma_{\text{soil}}\, z \qquad (4.5)$$

It will be seen later that the actual thermal expansion in integral bridge decks is closely comparable to that which occurs in a similar unrestrained deck (as the restraint offered by typical abutments and backfill is relatively small). Therefore, for a bridge deck of length, L, which experiences an increase in temperature of ΔT, d may be calculated as

$$d \approx \alpha \Delta T L/2 \qquad (4.6)$$

Implicit in Equations 4.3 and 4.4 is the assumption that a bank seat experiences a lateral translation, whereas a deeper abutment bends and rotates about a point just below the ground level on its inner face.

EXAMPLE 4.1   DETERMINATION OF MAXIMUM LATERAL ABUTMENT EARTH PRESSURES

A 50 m long integral bridge has deep wall abutments, which retain 6 m of well-compacted granular fill. The peak angle of friction of the fill is $\phi' = 45°$, and its dry density is 1900 kg/m³. The design extreme event for the determination of maximum abutment pressures is a 40° increase in temperature. The coefficient of thermal expansion for the deck is $\alpha = 12 \times 10^{-6}/°C$ and $\delta_a/\phi' = 0.67$. Find the maximum lateral abutment earth pressure.

From Equation 4.6

$$d = (12 \times 10^{-6})(40°)(50/2) = 0.012 \text{ m}$$

Figure 4.4 indicates that $K_p = 17.5$ for $\phi' = 45°$ and $\delta_a/\phi' = 0.67$. Equation 4.4 gives

$$K^* = \text{Min}\left[17.5, \left(\frac{0.012}{0.05 \times 6} \right)^{0.4} 17.5 \right] = 4.83$$

for $z \leq 0.66H$. The unit weight of the soil is

$$\gamma_{soil} = 1.9 \times 9.8 = 18.6 \text{ kN/m}^3$$

Therefore, for $z \leq 0.66H$ or $z \leq 4$ m,

$$\sigma_h(z) = (4.83)(18.6)z$$
$$\Rightarrow \sigma_h(z) = 359 \text{ kN/m}^2 \quad \text{at } z = 4 \text{ m.}$$

The at-rest lateral stress from Equation 4.5 is

$$\sigma_{h0}(z) = (1 - \sin 45°) \times 18.6z$$
$$\Rightarrow \sigma_{h0}(z) = 33 \text{ kN/m}^2 \quad \text{at } z = 6 \text{ m.}$$

As $\sigma_h(z)$ at $z = 4$ m exceeds $\sigma_{h0}(z)$ at $z = 6$ m, $\sigma_h(z)$ is taken as 359 kN/m² between $z = 4$ m and $z = 6$ m.

4.1.3 Stiffness of soil

The longitudinal expansion of integral bridge decks is resisted not just by the abutment supports but also by the backfill soil behind the abutments and the natural/imported soil beneath them. For most cases, it is necessary to quantify the restraint provided by the soil. This can only be achieved with a knowledge of the appropriate soil stiffness parameters. Clearly, higher soil stiffness will lead to higher axial forces and bending moments in the deck because of resistance to its longitudinal expansion or contraction. The design stiffness used for the calculation of such forces and moments should therefore be a maximum credible value.

Granular soil

It is now well established that cycling of granular soil behind an integral bridge abutment leads to a build-up or escalation of lateral earth stresses acting on the abutment. This phenomenon has also been referred to as 'soil ratcheting' (England et al. 2000) and leads to an increase in soil stiffness (and hence restraint to the deck) as cycling progresses. The stress–strain relationship for soil is non-linear at strains in excess of approximately 0.00005 ($= 50 \times 10^{-6}$), and predictions should therefore allow for this strain level dependency as well as for all of the other factors known to affect the stress–strain relationship of cyclically loaded soil. Uncertainties associated with such modelling are very high, and it is therefore preferable to employ an equivalent linear soil modulus (E_s) in calculations. This E_s value needs to be representative of the soil density and the average operational stress and strain levels in the soil mass under the cyclic conditions imposed by the abutment.

Lehane (2011) compiled a database of lateral stresses measured in physical modelling experiments involving integral bridge abutments with granular backfill and then used the finite element method to back-analyse the value of E_s after 100 cyclic rotations of an abutment, with a rotational amplitude (δ/H). It was found that E_s varies primarily with the initial vertical stress level (σ'_{v0}) and is not very sensitive to the initial backfill density at rotational amplitudes less than approximately 0.5%. This is essentially because looser soils increase in density under the action of low and moderate levels of cycling. The observed trends are analogous to the dominant effect of stress level on the resilient modulus employed in pavement engineering (e.g. see Indraratna et al. 2009). The database of lateral stress measurements showed that the escalation of lateral stresses on the abutment is effectively completed after 100 typical design cyclic rotations. At this stage, E_s can be described by the following empirical expression proposed by Lehane (2011) for the quoted range of rotational amplitudes, δ/H (where p_a is atmospheric pressure = 100 kPa):

$$\left.\frac{E_s}{p_a}\right|_{N=100} \approx [1000 \pm 300]\left(\frac{\sigma'_{v0}}{p_a}\right)^{0.7} \quad \text{for } 0.1\% < \frac{\delta}{H} < 0.5\% \tag{4.7}$$

Lehane (2011) also shows that the same value of E_s can be derived directly from the very small strain (truly elastic) modulus ($E_{\max}$) as

$$\left.E_s\right|_{N=100} = \frac{E_{\max}}{2.5} \quad \text{for } 0.1\% < \frac{\delta}{H} < 0.5\% \tag{4.8}$$

where $E_{\max}$ for the granular backfill can be readily measured directly using a variety of standard geophysical techniques that measure the shear wave velocity (V_s). The value of $E_{\max}$ in a granular soil with density ρ may be determined from the following expression:

$$E_{\max} = 2.2 \rho V_s^2 \tag{4.9}$$

If the rotational amplitude (δ/H) is less than 0.1%, the operational stiffness can be estimated as the truly elastic modulus, $E_{\max}$. If no shear wave velocity data are available, $E_{\max}$ can be approximated from Equations 4.7 and 4.8 so that

$$\left.E_s\right|_{N=100} = E_{\max} \approx 2500\left[\left(\sigma'_{v0}\right)^{0.7} p_a^{0.3}\right] \quad \text{for } \frac{\delta}{H} < 0.1\% \tag{4.10}$$

Clay soil

Lehane (2011) also showed that the ratcheting effect on stiffness due to cyclic loading is unlikely to be significant in clay soils. The value for E_s in clay can therefore be estimated using conventional empirical relationships with the clay's undrained shear strength (s_u). Bearing in mind typical strain amplitudes adjacent to integral bridge abutments, a maximum credible E_s value in clay may be estimated from

$$\left.E_s\right|_{N=100} = 500 \, s_u \quad \text{for } \frac{\delta}{H} < 0.5\% \tag{4.11}$$

Equation 4.11 can underestimate the operational value of E_s at very low strain levels. Such underestimation is, however, of little consequence as the stresses induced on the abutment at such strain levels are very small.

4.2 CONTRACTION OF BRIDGE DECK

There is generally a lesser height of soil in front of bridge abutments than behind them. As a result, the resistance provided by such soil to the contraction of a bridge deck is usually small. This means that, in an analysis to determine the effects of elastic shortening, creep and/or shrinkage, the principal uncertainty relates to the resistance to movement at the bases of the piers and abutments.

4.2.1 Contraction of bridge fully fixed at the supports

The case is first considered of an integral bridge in which no translational movement can occur at the base of the abutments. These conditions are applicable if the abutment foundations are cast in very dense soil or rock. However, an analysis of this type is often used as a first step to determine a limit on the stresses induced by deck contraction, when the supports are partially fixed.

The bridge illustrated in Figure 4.5a was considered in Chapter 3 (Example 3.5) for an axial contraction due to temperature of 20° in the deck (ABC). In that case, the bridge was fully restrained at the base of each abutment and pier. However, some movement of the deck was possible through bending in the abutments. If the ends of the deck, A and C, were fully prevented from contracting, the decrease in temperature would generate a large tensile force in the deck, and there would be no contraction. However, the resistance of the abutments to movement was considerably less than the axial stiffness of the deck.

The end result for this example was a relatively small axial tension in the deck, only 6.5% of the potential level for this temperature change, and a relatively large contraction. The axial contraction induced bending in the abutments and, due to the integral nature of the bridge, bending in the deck also. The complete bending moment diagram is illustrated in Figure 4.5b.

4.2.2 Contraction of bridge on flexible supports

Most bridges are constructed on supports which have some degree of flexibility. Abutments and piers are generally either supported on foundations bearing directly on the ground

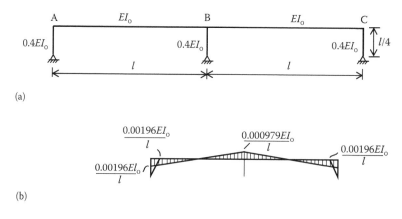

Figure 4.5 Contraction of frame pinned at supports: (a) geometry; (b) bending moment diagram from Example 3.5.

below or on pile caps underlain by piles. Quantification of the pile resistance is beyond the scope of this text, and interested readers are referred to books such as that of Tomlinson (1994). Strip foundations or pile caps are commonly found at around 0.5–1.0 m below the ground level on the inside of the abutment, as illustrated in Figure 4.6. It is this small depth of soil, together with sliding resistance at the base of the pad, that resists bridge contraction. The soil around the strip foundation can be idealised by a number of linear elastic springs.

Expressions for the stiffness of such springs have been deduced here from relationships provided by Dobry and Gazetas (1986) for an elastic soil. Design spring stiffnesses on the inside of the abutment for a strip foundation of width B, embedded to a depth of between 0.5 and 1.0 m below the ground level, are given in the following equation:

$$\left.\begin{array}{l} k_{vert} \approx 0.4E_s \\ k_{horz} \approx 0.5E_s \\ k_{rot} \approx \dfrac{E_s B^2}{6} \end{array}\right\} \tag{4.12}$$

where k_{vert}, k_{horz} and k_{rot} are the stiffnesses per metre length of strip foundation for vertical, horizontal and rotational displacement, respectively. Conservative, upper-bound estimates of the secant Young's modulus of elasticity, E_s, may be calculated using Equations 4.7 to 4.10.

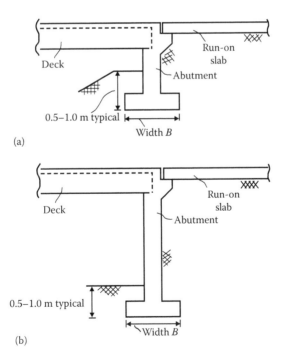

Figure 4.6 End of integral bridge showing shallow depth of soil on inside: (a) bank seat; (b) deep abutment.

EXAMPLE 4.2   CONTRACTION FOR SHALLOW STRIP FOUNDATION

The bridge illustrated in Figure 4.7 is subjected to a shrinkage strain of 200×10^{-6}. It is required to determine the distribution of bending moment and axial force generated in the deck, given that the Young's modulus for the concrete is 30×10^6 kN/m². The deck has a second moment of area of 0.07 m⁴, and a cross-sectional area of 0.6 m² per metre width. The wall has a second moment of area of 0.14 m⁴, and a cross-sectional area of 1.2 m² per metre width. The foundation is assumed to be on competent granular soil, working under a bearing pressure of 300 kN/m², and the breadth of the strip foundation is 2.5 m.

Equation 4.7 gives

$$\frac{E_s}{p_a} = [1000 \pm 300]\left(\frac{\sigma'_{v0}}{p_a}\right)^{0.7}$$

$$\Rightarrow \quad \frac{E_s}{100} = [1000 \pm 300]\left(\frac{300}{100}\right)^{0.7}$$

which gives a mean value of

$$E_s = (100)[1000]\left(\frac{300}{100}\right)^{0.7} = 216{,}000 \text{ kN/m}^2$$

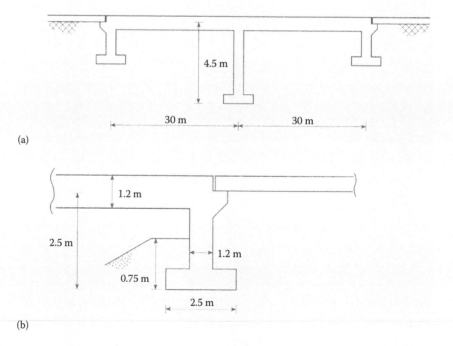

(a)

(b)

Figure 4.7 Bridge of Example 4.2: (a) elevation; (b) detail at abutment.

Equation 4.12 then gives spring stiffnesses per metre run for the supports of

$k_{vert} = 0.4E_s = 86,400$ kN/m/m

$k_{horz} = 0.5E_s = 108,000$ kN/m/m

$k_{rot} = E_sB^2/6 = 225,000$ kN/m/m

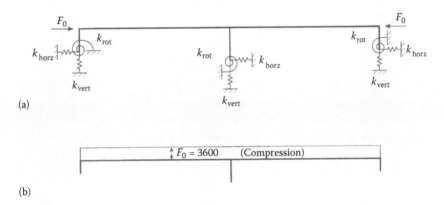

(a)

(b)

Figure 4.8 Computer model for bridge of Example 4.2: (a) equivalent loading and springs; (b) associated axial force diagram.

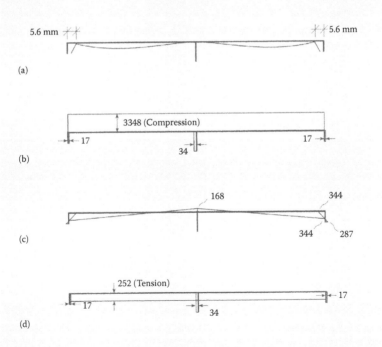

Figure 4.9 Analysis results: (a) deflected shape; (b) axial force diagram from computer analysis; (c) bending moment diagram; (d) corrected axial force diagram.

The equivalent load for a shrinkage strain of 200×10^{-6} is the product of the strain, the modulus of elasticity of concrete and the cross-sectional area (per metre run):

$$F_0 = (200 \times 10^{-6})(30 \times 10^6)(0.6)$$
$$= 3600 \text{ kN}$$

The equivalent loads and the associated axial force diagram are illustrated in Figure 4.8. The frame was analysed using a standard analysis package, which gave the deflected shape, axial force and bending moment diagrams illustrated in Figures 4.9a–c. Subtracting the associated axial force diagram gives the actual distribution of axial force generated by the shrinkage, illustrated in Figure 4.9d. No adjustment is necessary for the deflected shape or bending moment diagram.

Example 4.2 is interesting in that it gives an indication of the magnitude of bending moments and axial forces that can be generated by a restrained shrinkage. Out of a total potential shortening of 6 mm ($200 \times 10^{-6} \times 30{,}000$ mm) at each end, 5.6 mm is predicted to actually occur. However, the restraint that prevents the remaining 0.4 mm does generate distributions of stress in the frame. The axial tension is relatively small at 252 kN, corresponding to a stress in the deck of less than 0.4 N/mm². However, the bending moments at the ends are more significant at 344 kNm/m. Assuming uncracked conditions, this corresponds to a maximum flexural stress of ± 2.9 N/mm² if the deck were bending about its centre and more if the centroid is off-centre.

4.3 CONVENTIONAL SPRING MODEL FOR DECK EXPANSION

Soil generally provides considerably more resistance to deck expansion than contraction as abutments are backfilled up to the level of the underside of the run-on slab (Figure 4.6). Thus, the stresses generated by an increase in deck temperature, for example, will be affected significantly by the properties of the soil behind the abutments. The design value of E_s may be obtained from Equations 4.7 to 4.10. An approximate expression, assuming linear elasticity, has been developed for the horizontal spring stiffness per square metre of the backfill behind an abutment of depth H and transverse length, L_t:

$$k_{\text{horz}} \approx \frac{\left(\dfrac{4}{\pi}\right)E_s}{\left(\dfrac{L_t}{H}\right)^{0.6} H} \text{ kN/m/m}^2 \tag{4.13}$$

The application of Equation 4.13 is illustrated in the following example.

EXAMPLE 4.3   CONVENTIONAL SPRING MODEL

The culvert illustrated in Figure 4.10 is subjected to an increase in temperature of 20°C. The resulting distribution of bending moment is required, given that the culvert is made from concrete with an elastic modulus of 28×10^6 kN/m² and a coefficient of thermal expansion of 12×10^{-6}/°C. The bulk unit weight of the granular backfill is 20 kN/m³, and the piles are assumed to provide insignificant lateral restraint to the deck.

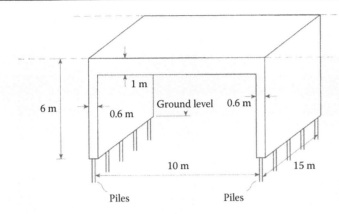

Figure 4.10 Culvert of Example 4.3.

To estimate the rotational amplitude induced in the backfill, the expansion of the culvert is estimated as its unrestrained value, that is, the product of the temperature increase, the coefficient of thermal expansion and the distance of the abutment from the stationary point (the centre of the culvert):

$$\delta = (20°)(12 \times 10^{-6})(5)$$
$$= 0.0012 \text{ m}$$

Hence, the approximate rotational amplitude is

$$\delta/H = (0.0012)/6 = 0.02\%$$

As this strain level is low and outside the range of applicability of Equation 4.7, Equation 4.10 is used to approximate the operational stiffness. Using the average initial vertical effective stress (σ'_{v0}) at the culvert (3×20 kN/m³ = 60 kN/m²), this gives

$$E_s \approx 2500[60^{0.7} \times 100^{0.3}]$$
$$\Rightarrow E_s = 175{,}000 \text{ kN/m}^2$$

The average horizontal spring stiffness is then given by Equation 4.13:

$$k_{horz} \approx \frac{\left(\dfrac{4}{\pi}\right)175{,}000}{\left(\dfrac{15}{6}\right)^{0.6}(6)} \text{ kN/m/m}^2$$
$$\Rightarrow k_{horz} = 21{,}500 \text{ kN/m/m}^2$$

The model for a 1 m strip of the frame is then as illustrated in Figure 4.11a. The equivalent loading is

$$F_0 = (20°)(12 \times 10^{-6})(28 \times 10^6)(1 \times 1)$$
$$= 6720 \text{ kN}$$

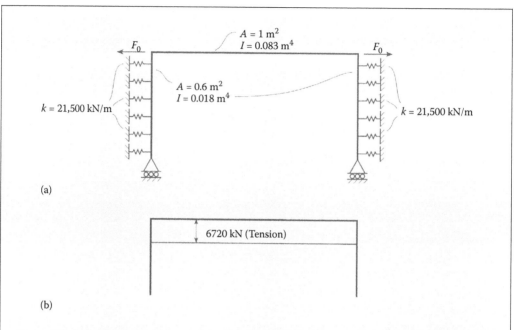

(a)

(b)

Figure 4.11 Computer model for culvert of Example 4.3: (a) springs and equivalent loads; (b) associated axial force diagram.

and the associated distribution of axial force is illustrated in Figure 4.11b. The bending moment diagram was found from a computer analysis and is illustrated in Figure 4.12. As there was no associated distribution of bending moment, this is the final distribution of moment due to the expansion. The moment in the abutments can be seen to change sign through its length due to the flexible nature of the horizontal support. The deflection found from the computer analysis was 1.18 mm, almost identical to the unrestrained deflection of $(20°)(12 \times 10^{-6})(5000) = 1.20$ mm.

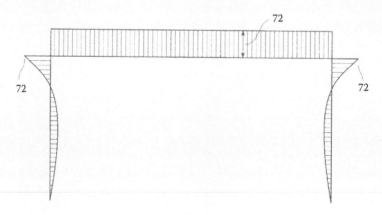

Figure 4.12 Bending moment diagram for Example 4.3.

4.4 MODELLING EXPANSION WITH AN
EQUIVALENT SPRING AT DECK LEVEL

An alternative to the conventional spring model is presented here, which has a number of advantages over the traditional approach. This technique consists of modelling both the abutment and the surrounding soil, with an equivalent lateral and rotational spring at deck level. The approach used to derive the spring constants represents the soil as a complete mesh of finite elements rather than a series of springs and is therefore considered to be sounder theoretically than the conventional spring model described in Section 4.3. This method does not, however, provide details concerning the distribution of moment in the abutment or the pressure distribution in the soil.

4.4.1 Development of general expression

Lehane (1999) and Lehane et al. (1999) determined the forces and moments associated with lateral displacement and rotation of the top of an abutment with retained backfill, that is, the forces and moments associated with passive movements, which occur as a consequence of deck expansion. They conducted a series of finite element analyses, which involved the application at the top of the abutment of (i) a horizontal translation δ with zero rotation and (ii) a rotation θ with zero horizontal translation (Figure 4.13).

The purpose of the analyses was to provide credible upper-bound estimates of soil resistance. It was therefore assumed conservatively that the soil had limitless compressive and tensile strength (e.g. no passive failure or abutment lifting was allowed) and that no slip between the abutment and the soil occurred (e.g. base sliding or slip on the abutment stem was not permitted). However, given that relatively small movements are required to reduce pressures to their minimum (active) values on the inner face of the abutment, the analyses assumed that any soil present on this side did not contribute to the resistance.

It was found that the flexural rigidity of the abutment (EI_a) and the ratio, r, defined as

$$r = E_s/(EI_a) \qquad (4.14)$$

were the most important factors controlling the magnitudes of the lateral force (F_h) and moment (M) at the top of the abutment (Figure 4.13). The values of F_h and M were also seen to increase systematically as the base width (B) increased and its height (H) reduced. Best-fit

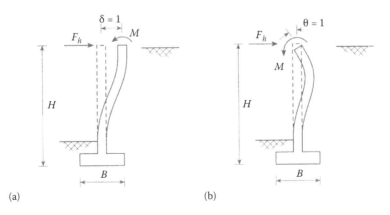

Figure 4.13 Stiffness components at the top of the abutment: (a) unit translation; (b) unit rotation.

Table 4.1 Range of parameters used in derivation of Equation 4.15

Parameter	Allowable range
E_s (kN/m^2)	10,000–500,000
EI_a (kNm^2/m)	1.0×10^4–2.5×10^6
$r = E_s/EI_a$ (m^{-4})	>0.05
H (m) (Figure 4.13)	1.5–12
B (m) (Figure 4.13)	0.5–3.5

expressions were obtained for F_h and M for the range of parameter values given in Table 4.1. They are given here in matrix form:

$$\begin{bmatrix} F_h \\ M \end{bmatrix} = EI_a \begin{bmatrix} f_1 r^{0.75} & f_2 r^{0.5} \\ f_2 r^{0.5} & r^{0.25} \end{bmatrix} \begin{bmatrix} \delta \\ \theta \end{bmatrix} \tag{4.15}$$

where f_1 and f_2 are functions of the ratio, H/B, which are given by Equation 4.16 for $r > 0.05$ m^{-4}. All values in this stiffness matrix can be reduced by 15% if friction between the abutment and soil is considered negligible.

$$\left. \begin{aligned} f_1 &= \frac{0.85 + \sqrt{H/B}}{3\sqrt{H/B}} \\ f_2 &= \frac{0.5 + \sqrt{H/B}}{2.5\sqrt{H/B}} \end{aligned} \right\} \tag{4.16}$$

For the range of parameters listed in Table 4.1, Equation 4.15 was found to predict values of F_h and M to within 10% of the values given by the finite element analyses. When a frame bridge with an abutment height of H is fixed rigidly at the supports and the system of fixities illustrated in Figure 4.14 is used, the stiffness matrix, $[K]$, in the absence of soil, is

$$[K] = \begin{bmatrix} \dfrac{EA_d}{L_d} + \dfrac{12EI_a}{H^3} & \dfrac{6EI_a}{H^2} & \cdots \\ \dfrac{6EI_a}{H^2} & \dfrac{4EI_a}{L_d} + \dfrac{4EI_a}{H} & \cdots \\ \vdots & \vdots & \ddots \end{bmatrix} \tag{4.17}$$

where A_d, L_d and I_d are the cross-sectional area, span length and second moment of area of the deck, respectively. When the bridge is embedded in soil and this is taken into account,

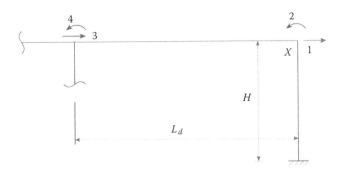

Figure 4.14 End part of frame bridge showing locations and directions of fixity.

the terms involving I_a and H are replaced with terms from Equation 4.15, with the result that Equation 4.17 becomes

$$[K'] = \begin{bmatrix} \dfrac{EA_d}{L_d} + EI_af_1r^{0.75} & EI_af_2r^{0.5} & \cdots \\[2ex] EI_af_2r^{0.5} & \dfrac{4EI_a}{L_d} + EI_ar^{0.25} & \cdots \\[2ex] \vdots & \vdots & \ddots \end{bmatrix} \qquad (4.18)$$

A comparison of Equation 4.17 and 4.18 shows that the influence of soil can be taken into account by analysing a model of a form similar to that illustrated in Figure 4.14. This could readily be achieved in analysis programs by allowing the appropriate stiffness terms to be changed in the program to those given in Equation 4.18. Alternatively, it is possible to allow for soil in a conventional structural analysis program through the use of an equivalent abutment second moment of area and height and the addition of a horizontal (translational) spring at X.

Equating the K_{22} (second row, second column) terms in Equations 4.17 and 4.18 gives

$$H_{eq} = \frac{4I_{eq}}{I_ar^{0.25}} \qquad (4.19)$$

where H_{eq} and I_{eq} are the equivalent abutment height and second moment of area, respectively. Similarly, equating the K_{12} (or K_{21}) terms gives

$$H_{eq} = \sqrt{\frac{6I_{eq}}{f_2I_ar^{0.5}}} \qquad (4.20)$$

Equations 4.19 and 4.20 can be simultaneously satisfied by selecting an equivalent abutment second moment of area of

$$I_{eq} = \frac{3I_a}{8f_2} \qquad (4.21)$$

The equivalent abutment height is then

$$H_{eq} = \frac{3}{2f_2} r^{-0.25} \tag{4.22}$$

To make the first terms (K_{11}) equal requires a further adjustment, which can be achieved by the addition of a linear horizontal spring at X of stiffness:

$$k_{sp} = EI_a r^{0.75} \left(f_1 - \frac{4f_2^2}{3} \right) \tag{4.23}$$

4.4.2 Expansion of frames with deep abutments

The equivalent single-spring model can be simplified for the case of deep abutments. For values of (H/B) in excess of 10, the parameters f_1 and f_2 approach their minimum values of 0.33 and 0.40, respectively. As a result, the equivalent abutment second moment of area can be set equal to the actual second moment of area without great loss of accuracy:

$$I_{eq} = I_a \tag{4.24}$$

Substituting for f_2 in Equation 4.22 gives an equivalent height of

$$H_{eq} = \frac{3.75}{r^{0.25}} \tag{4.25}$$

Finally, substituting for f_1 and f_2 in Equation 4.23 gives a spring stiffness (after rounding) of

$$k_{sp} = \frac{EI_a r^{0.75}}{8.5} \tag{4.26}$$

These equations can be used to estimate the properties of an equivalent frame for an integral bridge with deep abutments.

EXAMPLE 4.4   EQUIVALENT SINGLE-SPRING MODEL FOR FRAME WITH DEEP ABUTMENTS

The equivalent single-spring model is used to determine the maximum moment in the culvert illustrated in Figure 4.10 due to a temperature increase of 20°. The concrete has an elastic modulus of 28×10^6 kN/m^2 and a coefficient of thermal expansion of 12×10^{-6}/°C. The dry density of the backfill has been specified as 1600 kg/m^3.

The elastic modulus of the soil is found, as for Example 4.3, to be

$$E_s = 175{,}000 \text{ kN/m}^2$$

and the second moment of area of a 1 m strip of the abutment is

$I_a = (1)(0.6)^3/12 = 0.018 \text{ m}^4$

The ratio defined by Equation 4.14 is then

$r = E_s/EI_a$

$= \dfrac{175,000}{(28 \times 10^6)(0.018)}$

$= 0.35$

The equivalent height of abutment is then, from Equation 4.25,

$H_{eq} = 3.75/r^{0.25} = 3.75/0.35^{0.25} = 4.9 \text{ m}$

The stiffness of the single spring on each side is given by Equation 4.26

$k_{sp} = \dfrac{EI_a r^{0.75}}{8.5}$

$= \dfrac{(28 \times 10^6)(0.018)0.35^{0.75}}{8.5}$

$= 27,000 \text{ kN/m/m}$

The equivalent frame and loading are illustrated in Figure 4.15. The magnitude of the equivalent loads, as for Example 4.3, is

$F_0 = 6720 \text{ kN}$

The associated axial force diagram is as illustrated in Figure 4.11b. The model was analysed using a standard computer program, and the resulting bending moment diagram is illustrated in

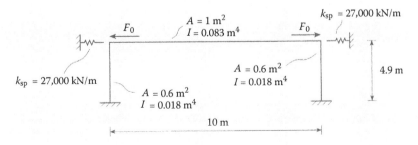

Figure 4.15 Computer model for bridge of Example 4.4.

Figure 4.16 Bending moment diagram from computer analysis of bridge of Example 4.4.

Figure 4.16. It is important to remember that the distribution of moment in the abutment is not realistic; the true shape of this distribution will be similar to that given in Figure 4.12. However, the magnitude of moment in the deck, 59 kNm, is likely to be more reliable than the value found in Example 4.3.

4.4.3 Expansion of bank-seat abutments

Equations 4.15 and 4.16 imply that an abutment provides a greater resistance to deck expansion if it has a lesser depth of embedment (H). This implication arises because of the assumption that the soil is an elastic material with infinite strength and that no sliding along the abutment base can take place. The reality, of course, is that shallow abutments are more likely to slide than deep ones and will therefore offer less restraint to deck expansion than Equations 4.15 and 4.16 would suggest.

The influence of a limited soil strength on the resistance offered by a bank seat is illustrated in Figure 4.17. In Figure 4.17b, predictions from finite element analyses are presented of a horizontal force/deflection relationship. When the soil is linear-elastic and infinitely strong, the function is, of course, linear. On the other hand, when the soil is treated as an elastic perfectly plastic material, with a finite strength defined by its friction angle, ϕ', the deflections per unit load can be seen to be significantly greater. Similar results can be shown for moment/rotation functions and for force/rotation and moment/deflection functions.

In the example of Figure 4.17, it can be seen that the effective lateral stiffness for a movement at the top of the abutment of 10 mm is only approximately half that of the purely elastic case. Effective rotational stiffnesses at this lateral movement are approximately 75% of the purely elastic case. It is not possible to generalise the observations made from calculations, such as those summarised in this figure, other than to say that the restraint provided by bank seats will be less than that predicted by Equation 4.15. It is therefore recommended that this equation be used in preliminary analysis and that a finite element soil/structure analysis incorporating a realistic constitutive model for the soil is performed if the effects of deck expansion have a significant influence on the final bridge design.

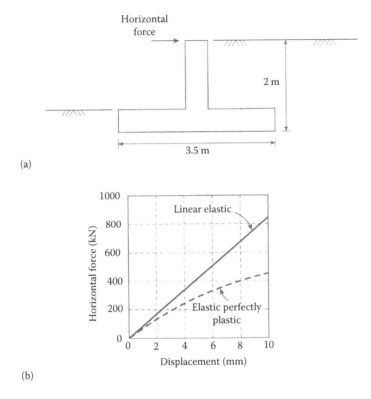

(a)

(b)

Figure 4.17 Finite element analysis results for bank seat abutment (E_s = 100,000 kN/m², E_c = 30 × 10⁶ kN/m², foundation bearing pressure = 200 kN/m², soil friction angle, ϕ' = 35°): (a) section through bank seat; (b) horizontal force/displacement relationship.

EXAMPLE 4.5 EQUIVALENT SINGLE-SPRING MODEL OF BANK SEAT

The equivalent single-spring model is used to determine the maximum moment in the culvert illustrated in Figure 4.18 due to a temperature increase of 20°C. The concrete has an elastic modulus of 28 × 10⁶ kN/m² and a coefficient of thermal expansion of 12 × 10⁻⁶/°C. The unit weight of the (loose) backfill is 16 kN/m³.

The elastic modulus for the soil is found in the same manner as for Example 4.3. Taking the average vertical effective stress, $\sigma'_{v0} = (2.5/2) \times 16 = 20$ kN/m³. Then taking the larger value possible in Equation 4.7 gives E_s = 42,000 kN/m². The second moment of area of a 1 m strip of the abutment is $I_a = 0.018$ m⁴. Hence, the ratio defined by Equation 4.14 is $r = 0.083$.

For this example, the ratio of embedment depth to foundation breadth, H/B, is 2.5/3 = 0.83. The parameters f_1 and f_2 are calculated from Equation 4.16:

$$f_1 = \frac{0.85 + \sqrt{H/B}}{3\sqrt{H/B}} = \frac{0.85 + \sqrt{0.83}}{3\sqrt{0.83}} = 0.64$$

$$f_2 = \frac{0.5 + \sqrt{H/B}}{2.5\sqrt{H/B}} = \frac{0.5 + \sqrt{0.83}}{2.5\sqrt{0.83}} = 0.62$$

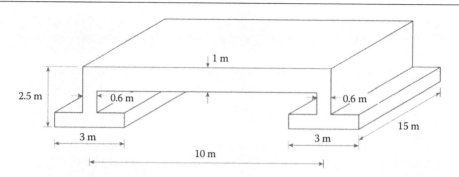

Figure 4.18 Bridge of Example 4.5.

The equivalent height is then calculated directly from Equation 4.22:

$$H_{eq} = \frac{3}{2f_2} r^{-0.25}$$

$$= \frac{3}{2(0.62)} 0.083^{-0.25}$$

$$= 4.5 \text{ m}$$

The equivalent abutment second moment of area is given by Equation 4.21:

$$I_{eq} = \frac{3I_a}{8f_2}$$

$$= \frac{3(0.018)}{8(0.62)}$$

$$= 0.011 \text{ m}^4$$

Finally, the spring stiffness is, from Equation 4.23,

$$k_{sp} = EI_a r^{0.75} \left(f_1 - \frac{4f_2^2}{3} \right)$$

$$= (28 \times 10^6)(0.018)0.083^{0.75} \left(0.64 - \frac{4(0.62)^2}{3} \right)$$

$$= 9900 \text{ kN/m/m}$$

The equivalent frame and loading are illustrated in Figure 4.19. The magnitude of the equivalent loads, as for Examples 4.3 and 4.4, is

$$F_0 = 6720 \text{ kN}$$

This model was analysed, and the bending moment diagram is illustrated in Figure 4.20. The maximum magnitude of moment in the deck due to the expansion is 69 kNm. It is clear from Figure 4.17 that this calculation is quite conservative.

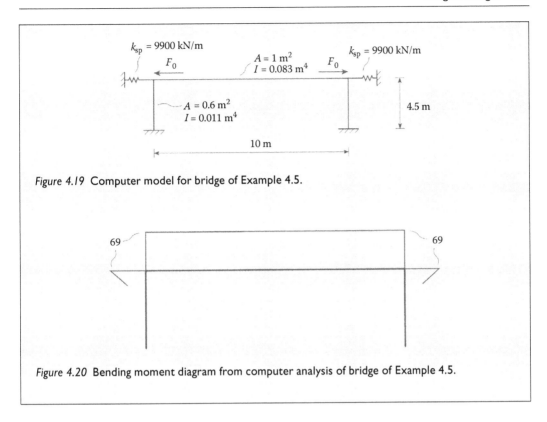

Figure 4.19 Computer model for bridge of Example 4.5.

Figure 4.20 Bending moment diagram from computer analysis of bridge of Example 4.5.

4.5 RUN-ON SLAB

It has been seen in this chapter that soil provides some restraint against deck movement in integral bridges but that most of the movement still takes place. On a road bridge, this must be accommodated if premature deterioration of the pavement is to be avoided. This is achieved in many cases by the installation of a run-on slab, as illustrated in Figure 4.21. The effect of such a slab is to allow relative rotation between the deck and the run-on slab

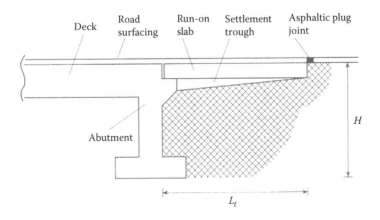

Figure 4.21 Run-on slab.

while preventing relative translation. Preventing relative vertical translation significantly improves the rideability for vehicles travelling over the bridge. Preventing relative horizontal translation is not so simple. Clearly, the bridge still expands and contracts relative to the surrounding soil, and the incorporation of a run-on slab does not prevent this. In effect, it transfers the relative horizontal movement from the end of the deck to the end of the run-on slab. This approach is widely adopted as the failure of a joint at the end of a run-on slab is a minor maintenance problem, whereas a leaking joint at the end of a deck can result in deterioration of the bridge itself.

Run-on slabs are designed to span the settlement troughs, assumed to be roughly triangular, that develop behind the abutments of integral bridges. An asphaltic plug joint positioned at the junction between the run-on slab and the bridge approach road is commonly used to facilitate horizontal movements. Settlement troughs arise because of the tendency for cohesionless backfill, whatever its density, to contract and increase in density in response to cyclic straining. Such straining is imposed on the backfill by the abutment, which moves in response to thermal movements of the deck. Analytical prediction of the shapes and magnitudes of settlement troughs is, however, not commonly attempted by bridge designers. This is because existing models, which attempt to simulate the soil's response to a complex history of cyclic straining, are very approximate, are difficult to use and require measurement of a large range of representative geotechnical parameters from cyclic laboratory tests.

Settlement profiles may be approximated as having a triangular shape, varying from a maximum settlement (δ_{max}) at the abutment to zero at a distance L_t from it. It has been shown by Springman et al. (1996) that, after many cycles of imposed lateral movement δ, δ_{max} varies between approximately 10δ and 20δ in well-compacted fill for both deep abutments and bank seats.

The assessment of the required length of the run-on slab relies on observations of measured behaviour and engineering judgement. Both analytical and model test studies have shown that the surface settlement trough tends to an equilibrium profile after a large number of cyclic abutment movements of the same magnitude. Much larger settlements occur in initially loose backfills where considerable volumetric contractions take place before an 'equilibrium' density is attained. The extent of the settlement trough is also controlled by the amount of backfill subjected to cyclic abutment movements and therefore, for a given movement of the top of the abutment, could be assumed to vary approximately with the height of the retained fill (H).

These observations and those taken during centrifuge model tests by Springman et al. (1996) suggest that the length of the trough (L_t) is unlikely to exceed the limits given in Table 4.2. As an example, the length of run-on slab required for the bridge of Example 4.5 (Figure 4.18) is calculated. As the backfill is loosely compacted (density = 1600 kg/m³) and the abutments are not deep, a maximum trough length of $2.1H$ can be assumed from Table 4.2. Hence, the run-on slab should have a length of $2.1(2.5) = 5.25$ m.

Table 4.2 Approximate upper limits on expected trough lengths

	Granular fill	
	Well compacted	Loosely compacted
Deep abutments	0.6H	1.4H
Bank seats	0.9H	2.1H

4.6 TIME-DEPENDENT EFFECTS IN COMPOSITE INTEGRAL BRIDGES

Many integral bridges are constructed using a combination of precast prestressed beams and in situ concrete, such as illustrated in Figure 4.22. When the in situ concrete is first poured, the precast beams are simply supported, and the self-weight of the bridge induces a sagging moment, as illustrated in Figure 4.23a. When the in situ concrete subsequently sets, the bridge acts as a frame, and imposed traffic loading generates sagging near the centres of the spans and hogging over the supports (Figure 4.23b). The net result is substantial sagging near the centres of the spans and some hogging over the supports (Figure 4.23c).

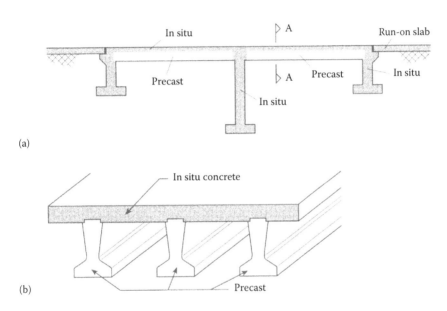

Figure 4.22 Composite integral bridge made from precast and in situ concrete: (a) elevation; (b) section A–A.

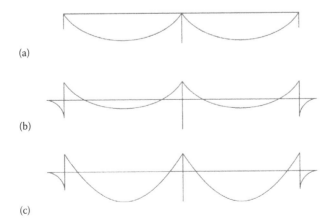

Figure 4.23 Bending moment diagrams due to short-term loading: (a) due to self-weight; (b) due to imposed traffic loading; (c) due to self-weight plus traffic loading (a plus b).

Non–prestressed reinforcement is generally provided at the top of the deck over the supports to resist the hogging moment, as illustrated in Figure 4.24, and it is often necessary to provide great quantities of closely spaced bars to prevent excessive cracking. Even before the bridge is made integral, there may be tension at points such as A in Figure 4.24, as these same pre-tensioned beams must be designed to resist substantial sagging moment near midspan. The resultant prestress force is therefore designed to be below the centroid near midspan (Figure 4.24) to ensure a hogging prestress moment. Near the supports, the hogging prestress moment combines with hogging due to applied loading, making it very difficult to prevent tension in the beams. The problem can be countered by the debonding of strands near the ends to prevent the prestress force from acting there.

All of the above effects occur in the short term, during the period immediately following the construction of the bridge. As outlined in Section 3.7, the distributions of bending moment change in the long term due to creep in the prestressed beams.

When prestress is first applied below the centroid, the beams hog upwards, as illustrated in Figure 4.25a. As they are simply supported, such curvature is unrestrained, so it results in instantaneous strain and a moment, which is the simple product of prestress force and eccentricity. Due to creep, these hogging strains increase with time. When the bridge is made integral, further curvature is resisted, and the resulting distribution of moment is as illustrated in Figure 4.25b. The long-term result is a distribution of prestress moment such as that illustrated in Figure 4.25c. This phenomenon is particularly significant if the bridge is made integral when the precast concrete is young, as this causes most of the creep strain to occur when it is in the integral form. It can result in cracking at the bottom of the deck over the supports, as illustrated in Figure 4.26, particularly at the interface between the precast and in situ concretes.

Example 3.18 demonstrates a method of analysing this kind of bridge for the effects of creep. A simpler approach is described by Clark and Sugie (1997), who carried out a parametric study of composite integral bridges. They suggest that there is little point in trying to determine the distribution of bending moment that develops in the long term, as there are few creep/shrinkage computer models that give consistently reliable results. In a study of continuous bridges made integral at the interior supports, they calculated the maximum long-term sagging moment for beams made integral when between 21 and 100 days old. They propose the assumption of a sagging moment of 750 kNm (per beam) for spans in the 20 to 36 m range where the beams are 1100 mm deep or greater. For smaller beams, they suggest designing for a moment of 600 kNm.

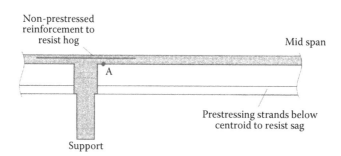

Figure 4.24 Detail near support of composite integral bridge.

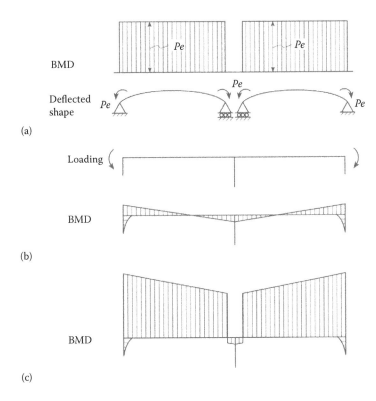

(a)

(b)

(c)

Figure 4.25 Effects of prestress on composite integral frame: (a) equivalent prestress loading and bending moment diagram at time of transfer of prestress; (b) equivalent prestress loading and bending moment diagram due to creep strains after frame is made integral; (c) total bending moment diagram due to prestress.

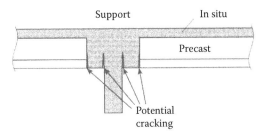

Figure 4.26 Detail at support showing points where long-term cracking is likely to occur.

Chapter 5

Slab bridge decks

Behaviour and modelling

5.1 INTRODUCTION

Bridges are often wide relative to their length and cannot be modelled accurately as beams or frames. Fortunately, slabs can be idealised using one of a number of well-proven methods and analysed using widely available analysis programs. To understand the basis of such programs and their limitations, it is necessary to first consider the theory of bending of plates.

5.2 THIN-PLATE THEORY

Slabs used in the construction of bridge decks are generally thin relative to their span lengths. Such slabs can be assumed to behave like thin plates, which can be thought of as the two-dimensional equivalent of beams. Thick plates correspond to deep beams and are not considered here. Thin plates get their strength from bending, in a similar way to beams, except that bending takes place in two mutually perpendicular directions in the plane of the plate.

5.2.1 Orthotropic and isotropic plates

A material in which the behaviour in one direction is independent of the other is referred to as *anisotropic*. A subset of anisotropic materials is *orthotropic* materials in which the behaviour varies in mutually perpendicular directions (X and Y) only. Orthotropy represents the most general material behaviour usually considered for bridge decks. A further subset of orthotropic materials is *isotropic* materials in which the behaviour in all directions is the same. Although this type of material is rarely found in bridge construction, isotropic plate theory can be used with reasonable accuracy for the analysis of many bridges.

A *materially* (or *naturally*) orthotropic plate is composed of a homogeneous material that has different elastic properties in two orthogonal directions but has the same geometric properties. This implies that the plate has a uniform thickness and hence, the same second moment of area in both directions but different moduli of elasticity. Such a plate might be constructed of a material where the microstructure is orientated in two mutually perpendicular directions such as timber. This type of plate is not typical of that found in bridge decks but is frequently used as an approximation of actual conditions.

Many bridge slabs possess different second moments of area in two directions such as reinforced concrete slabs with significantly different amounts of reinforcement in the two directions or voided slabs. These types of slab are referred to as *geometrically* (or *technically*) orthotropic. In the following sections, the theory of materially orthotropic thin plates is developed. While the theory is strictly only applicable to cases of material orthotropy, it is common practice to extend it to include geometric orthotropy. Thus, equations are derived, assuming the plate to have a uniform depth, but they are subsequently extended to decks which have different second moments of area in orthogonal directions.

5.2.2 Bending of materially orthotropic thin plates

Figure 5.1 shows a portion of a thin plate in the X–Y plane. The origin of the axis system is at mid-depth in the plate, at which point $z = 0$. Figure 5.2 shows a small segment of plate with dimensions $\delta x \times \delta y$ and a cube of material in that segment a distance z above the origin, which has a height of δz. In this figure, the thickness of the plate is taken to be d. When a load is applied, the cube both moves and distorts. Considering initially the X–Z plane, the points a, b, c and d shown in Figure 5.2 move to a', b', c' and d', as illustrated in Figure 5.3. The displacement of point a in the X direction is denoted u. Considering point b, a distance δx from a, the displacement at that point in the X direction, will be u plus the change in u over the distance δx:

$$\text{displacement at } b \text{ in } X\text{-direction} = u + \frac{\partial u}{\partial x}\delta x$$

Hence, the length of $a'b'$ projected onto the X axis is

$$\left. a'b' \right|_X = \left| ab \right| + \left(u + \frac{\partial u}{\partial x}\delta x \right) - (u)$$

$$\Rightarrow \quad \left. a'b' \right|_X = \delta x + \frac{\partial u}{\partial x}\delta x$$

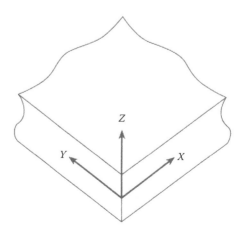

Figure 5.1 Portion of thin plate and coordinate axis system.

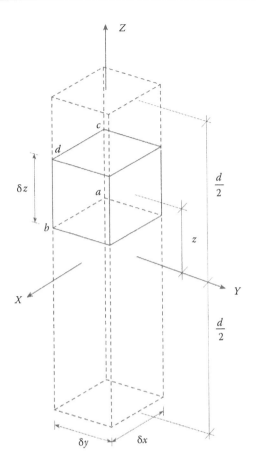

Figure 5.2 Segment of thin plate and elemental cube of material.

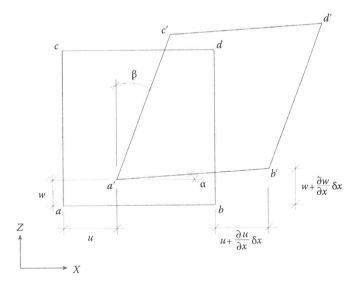

Figure 5.3 Distortion of cube of material in X–Z plane.

By definition, the strain in the X direction is

$$\varepsilon_x = \frac{|a'b'|_X - |ab|}{|ab|}$$

$$= \frac{\left(\delta x + \dfrac{\partial u}{\partial x}\delta x\right) - \delta x}{\delta x}$$

$$\Rightarrow \quad \varepsilon_x = \frac{\partial u}{\partial x} \tag{5.1}$$

Similarly, if v and w are the displacements in the Y and Z directions, respectively, it can be shown that

$$\varepsilon_z = \frac{\partial w}{\partial z} \tag{5.2}$$

and

$$\varepsilon_y = \frac{\partial v}{\partial y} \tag{5.3}$$

The shear strain in the X–Z plane is defined as the change in the angle cab from the original 90°, that is, the difference between $c'a'b'$ and cab. As can be seen in Figure 5.3, there are two components, α and β. Referring to Figure 5.3

$$\alpha = \frac{\left(w + \dfrac{\partial w}{\partial x}\delta x\right) - w}{\delta x + \left(u + \dfrac{\partial u}{\partial x}\delta x\right) - u}$$

As $\partial u/\partial x$ is small, this reduces to

$$\alpha = \frac{\partial w}{\partial x}$$

The other component of strain can be found similarly as

$$\beta = \frac{\partial u}{\partial z}$$

Hence, the shear strain is

$$\gamma_{xz} = \alpha + \beta$$

$$\Rightarrow \quad \gamma_{xz} = \frac{\partial u}{\partial z} + \frac{\partial w}{\partial x} \tag{5.4}$$

Similarly, the shear strains in the X–Y and Y–Z planes are, respectively,

$$\gamma_{xy} = \frac{\partial u}{\partial y} + \frac{\partial v}{\partial x} \tag{5.5}$$

$$\gamma_{yz} = \frac{\partial v}{\partial z} + \frac{\partial w}{\partial y} \tag{5.6}$$

In thin-plate theory, a number of assumptions are made to simplify the mathematics involved. The first of these assumptions is that there is no significant 'bearing' strain, that is, the strain in the Z direction is zero:

$$\varepsilon_z = \frac{\partial w}{\partial z} = 0 \tag{5.7}$$

This implies that w is independent of z, or that w is a function of x and y only. Figure 5.4 illustrates the implications of this assumption. The physical meaning is that there is no compression or extension of the bridge slab in a direction perpendicular to its plane. In other words, the depth of the slab remains unchanged throughout, and all points deflect vertically by exactly the same amount as the points directly above and below them. Clearly, this is a simplification, but the strains in the Z direction are generally so small that they have negligible effect on the overall behaviour of the bridge slab.

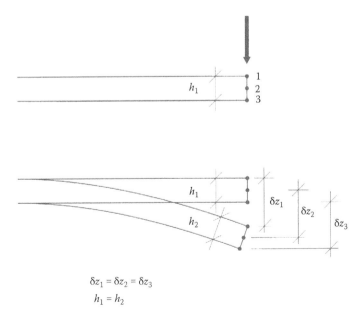

Figure 5.4 Segment of plate showing uniformity of distortion in Z direction.

The second assumption that is made is that the deflection of the plate is caused by bending alone and that shear distortion makes no significant contribution:

$$\gamma_{xz} = \frac{\partial u}{\partial z} + \frac{\partial w}{\partial x} = 0 \qquad (5.8)$$

and

$$\gamma_{yz} = \frac{\partial v}{\partial z} + \frac{\partial w}{\partial y} = 0 \qquad (5.9)$$

The consequences of Equation 5.8 are shown in Figure 5.5. The equation requires $\partial u/\partial z$ and $\partial w/\partial x$ to be equal and opposite. It follows from Figure 5.5 that the 90° angle of *cab* is preserved in the distorted *c'a'b'*. This assumption is again a simplification of the true behaviour but is justified by the fact that, with bridge slabs being relatively thin, their behaviour is dominated by bending rather than shear deformation. Notwithstanding this, concrete bridge slabs do not have great shear strength, and although shear strains are small, a means for determining shear stresses will be required. Such a method is presented later in this section.

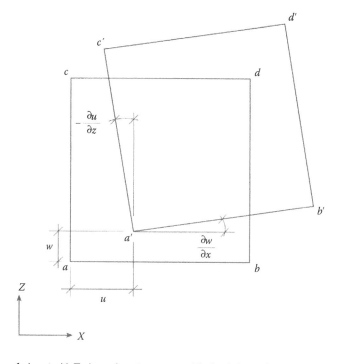

Figure 5.5 Segment of plate in *X–Z* plane showing assumed lack of shear distortion.

Rearranging Equation 5.8 gives

$$\frac{\partial u}{\partial z} = -\frac{\partial w}{\partial x}$$

$$\Rightarrow \quad u = \int\left(-\frac{\partial w}{\partial x}\right)dz$$

As w is independent of z, this implies

$$u = z\left(-\frac{\partial w}{\partial x}\right) + C \tag{5.10}$$

where C is a constant of integration. As the origin is located at the centre of the plate and bending is assumed to occur about that point, there is no displacement in either the X or Y directions at $z = 0$. Hence, at $z = 0$, u and v are both zero. Substituting this into Equation 5.10 implies that the constant C is zero, giving

$$u = -z\frac{\partial w}{\partial x} \tag{5.11}$$

By rearranging Equation 5.9, a similar expression can be derived for v:

$$v = -z\frac{\partial w}{\partial y} \tag{5.12}$$

Substituting Equations 5.11 and 5.12 into Equations 5.1 and 5.3, respectively, gives

$$\varepsilon_x = -z\frac{\partial^2 w}{\partial x^2} \tag{5.13}$$

$$\varepsilon_y = -z\frac{\partial^2 w}{\partial y^2} \tag{5.14}$$

Similarly, Equation 5.5 gives

$$\gamma_{xy} = -2z\frac{\partial^2 w}{\partial x \partial y} \tag{5.15}$$

In the flexural theory of beams, the curvature is defined as

$$\kappa = \frac{1}{R} \approx \frac{\partial^2 w}{\partial x^2}$$

where κ is the curvature and R is the radius of curvature. In the thin-plate theory, the equations are similar, but there are now curvatures in the X, Y and XY directions, which are given by

$$\kappa_x = \frac{\partial^2 w}{\partial x^2} \tag{5.16}$$

$$\kappa_y = \frac{\partial^2 w}{\partial y^2} \tag{5.17}$$

$$\kappa_{xy} = \frac{\partial^2 w}{\partial x \partial y} \tag{5.18}$$

Substituting Equations 5.16–5.18 into Equations 5.13–5.15, respectively, then gives

$$\varepsilon_x = -z\kappa_x \tag{5.19}$$

$$\varepsilon_y = -z\kappa_y \tag{5.20}$$

$$\gamma_{xy} = -2z\kappa_{xy} \tag{5.21}$$

Examination of Equation 5.19 shows that strain in the X direction is a linear function of z, as $\kappa_x = \partial^2 w/\partial x^2$ is independent of z; the strain distribution is triangular, as in beam theory. Equation 5.20 shows that the same applies to the strain in the Y direction. From this, it follows that plane sections remain plane as is assumed in beam theory. This is generally a reasonable assumption for slab bridge decks, but some cases do exist where this is not so. Such cases are discussed further in Chapter 7.

5.2.3 Stress in materially orthotropic thin plates

In the previous section, expressions were established for the various strains in a thin plate. Expressions are now developed for the corresponding stresses. Figure 5.6a shows a one-dimensional bar subjected to a tensile force. The only significant strain in this system is in a direction parallel to the axis of the bar. This strain, ε, is related to the stress, σ, and modulus of elasticity, E, by

$$\varepsilon = \frac{\sigma}{E}$$

In the three-dimensional case, strains in the other two directions become significant, as shown in Figure 5.6b. By defining the X axis as the direction of the applied force, the strain in that direction is given by

$$\varepsilon_x = \frac{\sigma_x}{E_x} - \frac{v_y \sigma_y}{E_y} - \frac{v_z \sigma_z}{E_z} \tag{5.22}$$

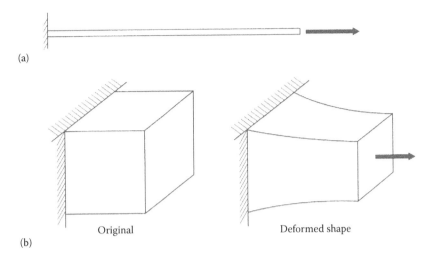

(a)

Original   Deformed shape

(b)

Figure 5.6 Distortion in one and three dimensions: (a) one-dimensional bar; (b) three-dimensional body showing the effect of stress in the axial direction on strains in the orthogonal directions.

where E_x, E_y and E_z are the moduli of elasticity in the X, Y and Z directions, respectively, and v_x, v_y and v_z are the corresponding Poisson's ratios. Equation 5.22 assumes that the plate is made of a homogenous material, and that the elastic constants (E_x, v_x, etc.) are independent of each other, as is appropriate for the materially orthotropic (or anisotropic) case.

For a thin plate in bending, the stress in the Z direction is small, and the Poisson's ratio is generally small for bridge deck materials. Consequently, the last term of Equation 5.22 can be ignored. An expression for strain in the X direction for the case of an orthotropic material with the elastic constants varying in the X and Y directions is then given by

$$\varepsilon_x = \frac{\sigma_x}{E_x} - \frac{v_y \sigma_y}{E_y}$$

and likewise, the strain in the Y direction is given by

$$\varepsilon_y = \frac{\sigma_y}{E_y} - \frac{v_x \sigma_x}{E_x}$$

In matrix format, this becomes

$$\left\{ \begin{array}{c} \varepsilon_x \\ \varepsilon_y \end{array} \right\} = \left[\begin{array}{cc} \dfrac{1}{E_x} & \dfrac{-v_y}{E_y} \\ \dfrac{-v_x}{E_x} & \dfrac{1}{E_y} \end{array} \right] \left\{ \begin{array}{c} \sigma_x \\ \sigma_y \end{array} \right\}$$

and by rearranging and inverting the matrix, we get

$$\left\{\begin{array}{c} \sigma_x \\ \sigma_y \end{array}\right\} = \frac{E_x E_y}{1 - \nu_x \nu_y} \begin{bmatrix} \dfrac{1}{E_y} & \dfrac{\nu_y}{E_y} \\ \dfrac{\nu_x}{E_x} & \dfrac{1}{E_x} \end{bmatrix} \left\{\begin{array}{c} \varepsilon_x \\ \varepsilon_y \end{array}\right\}$$

which yields expressions for the stresses as follows:

$$\sigma_x = \frac{E_x}{1 - \nu_x \nu_y}(\varepsilon_x + \nu_y \varepsilon_y) \tag{5.23}$$

$$\sigma_y = \frac{E_y}{1 - \nu_x \nu_y}(\varepsilon_y + \nu_x \varepsilon_x) \tag{5.24}$$

The shear modulus, G_{xy}, is defined as the ratio of shear stress, τ_{xy} to shear strain, γ_{xy}, which gives

$$\tau_{xy} = G_{xy}\gamma_{xy} \tag{5.25}$$

Substituting Equations 5.19–5.21 into Equations 5.23–5.25, respectively, gives expressions for the stresses in terms of curvature:

$$\sigma_x = \frac{-zE_x}{1 - \nu_x \nu_y}(\kappa_x + \nu_y \kappa_y) \tag{5.26}$$

$$\sigma_y = \frac{-zE_y}{1 - \nu_x \nu_y}(\kappa_y + \nu_x \kappa_x) \tag{5.27}$$

$$\tau_{xy} = -2zG_{xy}\kappa_{xy} \tag{5.28}$$

5.2.4 Moments in materially orthotropic thin plates

Figure 5.7 shows a small cube taken from a thin plate with the associated normal stresses and shear stresses. It is well established that, to satisfy equilibrium, pairs of shear stresses must be equal as follows:

$$\left.\begin{array}{c} \tau_{xy} = \tau_{yx} \\ \tau_{xz} = \tau_{zx} \\ \tau_{yz} = \tau_{zy} \end{array}\right\} \tag{5.29}$$

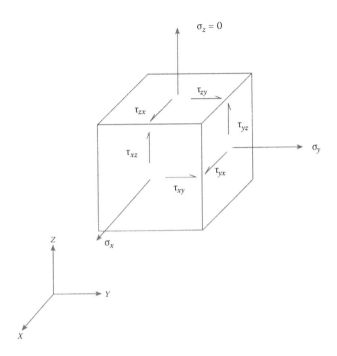

Figure 5.7 Elemental cube of material showing normal and shear stresses.

Considering the normal stresses first, Figure 5.8a shows a vertical line of cubes (such as that of Figure 5.7) through the depth of the plate in the X–Z plane. Each of these cubes is subjected to a normal stress in the X direction as indicated in the figure. When there are no in-plane forces in a bridge deck, the sum of the forces in these cubes is zero. As each cube is of the same surface area, it follows that

$$\sum_{i=1}^{n} \sigma_{xi} = 0$$

However, there is a bending moment caused by these stresses. The term m_x is used to represent the moment per unit breadth due to the σ_x stresses, summed through the depth of the deck. Figure 5.8b shows the depths of the cubes δz and their distances from the origin, z_1, z_2, z_3 and so on. Each cube has a width perpendicular to the page of δy (not shown in the figure). The forces F_1, F_2, F_3 and so on due to each of the stresses are shown. The ith cube contributes a component of hogging bending moment of magnitude $(\sigma_{xi}\delta z\delta y)z_i$. Taking sagging moment as positive and summing over the depth of the plate gives

$$m_x dy = - \int_{-d/2}^{d/2} (\sigma_x \, dz \, dy)z$$

$$\Rightarrow \quad m_x = - \int_{-d/2}^{d/2} \sigma_x z \, dz \qquad (5.30)$$

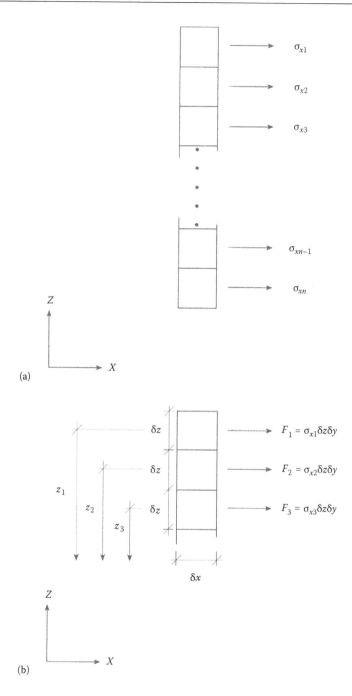

Figure 5.8 Vertical line of elemental cubes through the depth of a plate: (a) stresses on each cube; (b) forces on the cubes and distances from the origin.

Substituting Equation 5.26 into Equation 5.30 gives

$$m_x = \int_{-d/2}^{d/2} \left(\frac{z^2 E_x}{1 - \nu_x \nu_y} \right) (\kappa_x + \nu_y \kappa_y) \, dz$$

which gives

$$m_x = \frac{E_x}{1 - \nu_x \nu_y} (\kappa_x + \nu_y \kappa_y) \int_{-d/2}^{d/2} z^2 \, dz$$

$$\Rightarrow \quad m_x = \frac{E_x}{1 - \nu_x \nu_y} (\kappa_x + \nu_y \kappa_y) \left(\frac{z^3}{3} \right)_{-d/2}^{d/2}$$

$$\Rightarrow \quad m_x = \frac{E_x d^3}{12(1 - \nu_x \nu_y)} (\kappa_x + \nu_y \kappa_y) \tag{5.31}$$

Applying a similar method, it can be shown that the stress σ_y causes a moment per unit breadth m_y, which is given by

$$m_y = \frac{E_y d^3}{12(1 - \nu_x \nu_y)} (\kappa_y + \nu_x \kappa_x) \tag{5.32}$$

The second moment of area per unit breadth of the plate, i, is defined by

$$i = \int_{-d/2}^{d/2} z^2 \, dz = \frac{d^3}{12} \tag{5.33}$$

Therefore, Equations 5.31 and 5.32 can be rewritten in terms of the second moment of area as follows:

$$m_x = \frac{E_x i}{(1 - \nu_x \nu_y)} (\kappa_x + \nu_y \kappa_y) \tag{5.34}$$

$$m_y = \frac{E_y i}{(1 - \nu_x \nu_y)} (\kappa_y + \nu_x \kappa_x) \tag{5.35}$$

It is important to remember that m_x is the moment per unit breadth on a face perpendicular to the X axis and not about the X axis; in a reinforced concrete deck, it is the moment that would be resisted by reinforcement parallel to the X axis. Likewise, m_y is the moment per unit breadth on a face perpendicular to the Y axis.

Referring to Figure 5.7, it can be seen that the shear stresses result in forces parallel to the Y axis, which can also cause a moment. The moment per unit breadth due to τ_{xy} is termed m_{xy}. Figure 5.9 shows a number of cubes through the depth of the plate in the Y–Z plane. The shear force on the face of each cube is given by

$$\text{shear force} = \tau_{xy}\delta z\delta y$$

and the moment per unit breadth due to this force is given by

$$\text{moment per unit breadth} = \frac{(\tau_{xy}\delta z\delta y)z}{\delta y}$$

Taking anti-clockwise as positive on the $+X$ face, the total moment per unit breadth due to τ_{xy} is given by

$$m_{xy} = -\int_{-d/2}^{d/2} \tau_{xy} z \, \mathrm{d}z \qquad (5.36)$$

Substituting Equation 5.28 into Equation 5.36 gives

$$m_{xy} = -\int_{-d/2}^{d/2} (-2zG_{xy}\kappa_{xy})z \, \mathrm{d}z$$

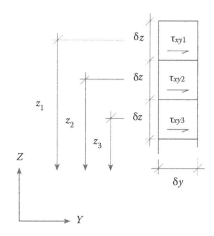

Figure 5.9 Stack of elemental cubes in the Y–Z plane showing shear stresses.

which gives

$$m_{xy} = 2G_{xy}\kappa_{xy} \int_{-d/2}^{d/2} z^2\,dz$$

$$\Rightarrow\ m_{xy} = \frac{G_{xy}d^3}{6}\kappa_{xy} \tag{5.37}$$

Similarly, the moment per unit length, m_{yx}, caused by τ_{yx} (on the Y face) can be shown to be

$$m_{yx} = -\int_{-d/2}^{d/2} \tau_{yx}z\,dz \tag{5.38}$$

$$\Rightarrow\ m_{yx} = \frac{G_{xy}d^3}{6}\kappa_{yx} \tag{5.39}$$

However, as indicated in Equation 5.29, equilibrium requires τ_{xy} and τ_{yx} to be equal, and comparison of Equations 5.36 and 5.38 yields

$$m_{xy} = m_{yx} \tag{5.40}$$

It follows from the definition of curvature (Equation 5.18) that the two twisting curvatures are the same:

$$\kappa_{xy} = \kappa_{yx} \tag{5.41}$$

so that there is no contradiction between Equations 5.37 and 5.39. These equations can be rewritten as

$$m_{xy} = m_{yx} = G_{xy}j/\kappa_{xy} \tag{5.42}$$

where j is known as the torsional constant per unit breadth and is given by

$$j = \frac{d^3}{6} \tag{5.43}$$

The moment m_{xy} ($= m_{yx}$) is often referred to as a twisting moment and is distinct from the normal moments m_x and m_y. Figure 5.10a shows the direction in which each of these moments acts, whereas Figure 5.10b shows the type of deformation associated with each of them.

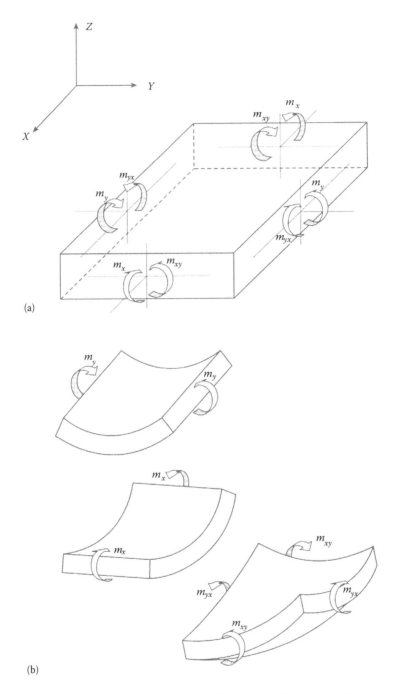

(a)

(b)

Figure 5.10 Bending and twisting moments in a plate: (a) segment of plate and directions of moments; (b) associated distortions.

5.2.5 Shear in thin plates

Vertical shear forces occur in bridge decks due to the shear stresses, τ_{xz} ($= \tau_{zx}$) and τ_{yz} ($= \tau_{zy}$) illustrated in Figure 5.7. Unlike beams, there are two shear forces at each point, one for each direction (X and Y). Defining q_x and q_y as the downward shear forces per unit breadth on the positive X and Y faces, respectively, then gives

$$q_x = -\int_{X \text{ face}} \tau_{xz} \, dz$$

and

$$q_y = -\int_{Y \text{ face}} \tau_{yz} \, dz$$

It was assumed earlier (Equations 5.8 and 5.9) that shear deformations in the plate were negligible. This is a reasonable assumption as shear deformation is generally small in bridge slabs relative to bending deformation. However, shear *stresses*, while numerically small, can be significant, particularly in concrete slabs, which are quite weak in shear. In the simple flexural theory of beams, the same phenomenon exists, and an expression is found from equilibrium of forces on a segment.

Figure 5.11 shows a segment of a beam of length dx in bending. The moment and shear force at the left end are M and Q, respectively, and at the right end are $M + dM$ and $Q + dQ$, respectively. Taking moments about the left-hand end gives

$$-M + (M + dM) = (Q + dQ)dx$$

Rearranging and ignoring the term, $dQdx$, which is relatively small, gives an expression for the shear force Q:

$$Q = \frac{dM}{dx} \tag{5.44}$$

In other words, the shear force is the derivative of the moment. In thin-plate theory, a similar expression can be derived.

A small element from the plate of base dimensions $dx \times dy$ is shown in Figure 5.12, with varying bending moment and shear force. The terms q_x and q_y refer to shear forces per unit

Figure 5.11 Equilibrium of small segment of beam.

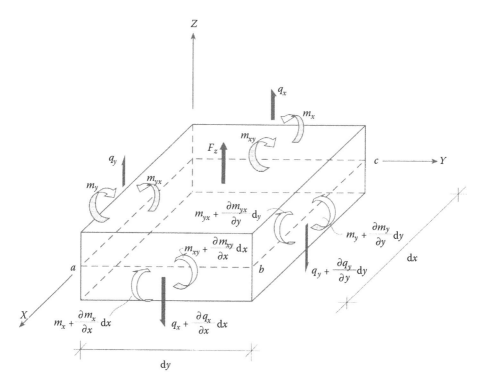

Figure 5.12 Equilibrium of small segment of slab.

breadth, while m_x, m_y and m_{xy} refer to moments per unit breadth. (This is different from the beam example above where Q and M referred to total shear force and total moment.)

Taking moments about the line a–b (Figure 5.12) gives

$$
-q_x \mathrm{d}y\ \mathrm{d}x - q_y \mathrm{d}x \frac{\mathrm{d}x}{2} + \left(q_y + \frac{\partial q_y}{\partial y}\mathrm{d}y\right)\frac{\mathrm{d}x^2}{2} - F_z \frac{\mathrm{d}x}{2} - m_x \mathrm{d}y
$$

$$
+ \left(m_x + \frac{\partial m_x}{\partial x}\mathrm{d}x\right)\mathrm{d}y - m_{yx}\mathrm{d}x + \left(m_{yx} + \frac{\partial m_{yx}}{\partial y}\mathrm{d}y\right)\mathrm{d}x = 0
$$

where F_z is the body force acting on the segment of slab (e.g. gravity). Dividing across by $\mathrm{d}x\mathrm{d}y$ gives

$$
-q_x + \frac{\partial q_y}{\partial y}\frac{\mathrm{d}x}{2} - f_z \frac{\mathrm{d}x}{2} + \frac{\partial m_x}{\partial x} + \frac{\partial m_{yx}}{\partial y} = 0
$$

where f_z is the body force per unit area. The second and third terms of this equation represent very small quantities and can be ignored, giving

$$
q_x = \frac{\partial m_x}{\partial x} + \frac{\partial m_{yx}}{\partial y} \tag{5.45}
$$

By taking moments about the line b–c (Figure 5.12), an equation for q_y can be derived in a similar manner:

$$q_y = \frac{\partial m_y}{\partial y} + \frac{\partial m_{xy}}{\partial x} \tag{5.46}$$

It can be seen that the expressions for the shear forces per unit breadth (Equations 5.45 and 5.46) are of a similar form to that for a beam (Equation 5.44), except for the addition of the last term involving the derivative of m_{xy} or m_{yx}.

5.3 GRILLAGE ANALYSIS OF SLAB DECKS

The idea of grillage analysis has been around for some time, but the method only became practical with the increased availability of computers in the 1960s. Finite element (FE) analysis software is more readily available today, but grillage models are still widely used for bridge deck analysis. Some of the benefits that have been quoted are that grillage analysis is easy to use and comprehend. These benefits traditionally favoured the method over FE analysis, which was typically only used for more complex problems. In today's environment of inexpensive, high-powered computers, coupled with elaborate analysis programs and user-friendly graphical interfaces, the FE method is replacing the grillage method, even for more straightforward bridge decks.

The plane grillage method involves the modelling of a bridge slab as a skeletal structure made up of a mesh of beams lying in one plane. Figure 5.13a shows a simple slab bridge deck supported on a number of discrete bearings at each end, and Figure 5.13b shows an equivalent grillage mesh. Each grillage member represents a portion of the slab, with the

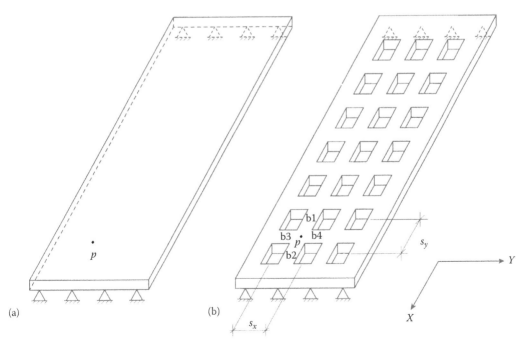

Figure 5.13 Grillage idealisation of a slab: (a) original slab; (b) corresponding grillage mesh.

longitudinal beams representing the longitudinal stiffness of that part of the slab, and the transverse grillage members representing the transverse stiffness. In this way, the total stiffness of any portion of the slab is represented by two grillage members. The grillage mesh and individual beam properties are chosen with reference to the part of the slab that they represent. The aim is that deflections, moments and shears will be the same in the slab and the grillage model. As the grillage is only an approximation, this will never be achieved exactly.

5.3.1 Similitude between grillage and bridge slab

It is necessary to achieve correspondence or *similitude* between the grillage model and the corresponding bridge slab. A point p is illustrated in Figure 5.13 corresponding to the junction of longitudinal beams, b1 and b2, and transverse beams, b3 and b4. Figure 5.14 shows an enlarged view of the junction along with the forces and moments acting on beams b1 and b3 in the grillage. The forces and moments have not been shown on beams b2 and b4 for clarity. The moments at the ends of beams b1 and b2 adjacent to p in the grillage give a measure of the moment per unit breadth m_x in the slab, whereas the moments at the ends of beams b3 and b4 give a measure of m_y.

The moments in the grillage members are total moments, whereas those that are required in the slab are moments per unit breadth. Therefore, it is necessary to divide the grillage member moments by the breadth of slab represented by each. This breadth is indicated in Figure 5.13 as s_x and s_y for the longitudinal and transverse beams, respectively. Unfortunately, in the grillage, the moments at the ends of beams b1 and b2 adjacent to p are generally not equal, nor are those in beams b3 and b4. For a fine grillage mesh, the difference is generally small, and it is sufficiently accurate to take the average moment at the ends of the beams

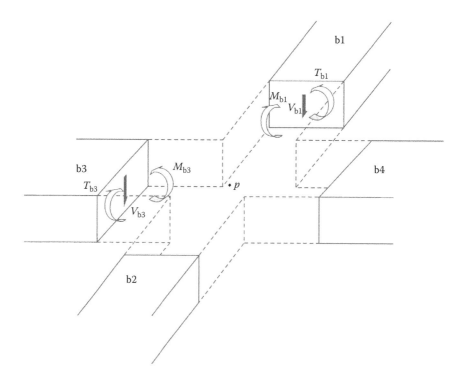

Figure 5.14 Segment of grillage mesh showing forces and moments on members b1 and b3.

meeting at the junction. The magnitude of this difference is often used as a check on the accuracy of the grillage, but it should be borne in mind that a small inequality does not necessarily mean an accurate grillage, as other factors may be involved.

The moments per unit breadth in the slab at point p are therefore obtained from the grillage using the following equations, with reference to Figures 5.13 and 5.14:

$$m_x \approx \frac{M_{b1}}{s_x} \approx \frac{M_{b2}}{s_x}$$

or

$$m_x = \frac{M_{b1} + M_{b2}}{2s_x}$$

Similarly,

$$m_y = \frac{M_{b3} + M_{b4}}{2s_y}$$

The moments at any other point in the slab can be found in a similar way. If the point is not at the intersection of longitudinal and transverse grillage members, it is necessary to interpolate between adjacent beams. Care should be taken while doing this, especially if a coarse grillage mesh is used. Some computer programs carry out this interpolation automatically, in which case it is necessary to confirm that the program has interpolated the results in a sensible manner. It is often more convenient to start by considering the locations at which moments will be required, and to formulate the grillage mesh in such a way as to avoid the need for interpolation between beams.

The twisting moments per unit breadth in the slab, m_{xy} and m_{yx}, are found from the torques in the grillage members in a similar manner. These moments at point p (again with reference to Figures 5.13 and 5.14) are given by

$$m_{xy} = \frac{T_{b1} + T_{b2}}{2s_x}$$

and

$$m_{yx} = \frac{T_{b3} + T_{b4}}{2s_y}$$

Equation 5.40 states that m_{xy} and m_{yx} are equal for materially orthotropic plates, but the torques in grillage members b1 and b2 will not necessarily be equal to the torques in b3 and b4. Therefore, the twisting moment in the slab is arrived at by averaging the torques per unit breadth in all four beams meeting at the point p. This may be quite unsatisfactory, as large variations of torque may exist between the longitudinal and transverse beams, particularly for orthotropic plates with significantly different flexural stiffnesses in the two directions. The situation can be improved by choosing torsion constants for the longitudinal and transverse beams, which promote similar levels of torque per unit breadth in both. This technique is discussed further in Section 5.3.2.

The shear forces per unit breadth in the slab, q_x and q_y, are found from the shear forces in the grillage members in a similar manner to the moments. At point p (Figures 5.13 and 5.14), these are given by

$$q_x = \frac{V_{b1} + V_{b2}}{2s_x}$$

and

$$q_y = \frac{V_{b3} + V_{b4}}{2s_y}$$

Equations 5.45 and 5.46 gave expressions for the shear forces per unit breadth in the slab. Examining, for example, the shear force V_{b1} in Figure 5.14, it can be seen that this shear force will be equal to the derivative of the moment M_{b1} with respect to x as this beam will comply with Equation 5.44. This accounts for the first term of Equation 5.45, but there is no account taken in the grillage analysis of the second term, namely, the derivative of m_{yx} with respect to y. This could be calculated in the grillage by finding the derivative of the torques in b3 and b4 with respect to y. However, unless m_{yx} is particularly large, this is not normally done as the resulting inaccuracy in the shear forces tends to be small.

5.3.2 Grillage member properties: Isotropic slabs

A grillage member in bending behaves according to the well-known flexure formula:

$$\frac{M}{I} = \frac{E}{R}$$

where M is the moment, I is the second moment of area, E is the modulus of elasticity and R is the radius of curvature. By substituting the curvature $1/R$ with κ and rearranging, the moment per unit breadth m is found:

$$m = Ei\kappa \tag{5.47}$$

where i is the second moment of area per unit breadth. Equation 5.34 gives an expression for the moment per unit breadth on the X face of the slab. For an isotropic slab, there is only one value for E and v. Substituting E for E_x and v for v_x and v_y in that equation gives

$$m_x = \frac{Ei}{(1 - v^2)}(\kappa_x + v\kappa_y)$$

As Poisson's ratio v is relatively small in bridge slabs (approximately 0.2 for concrete), it is common practice to ignore the second term in this equation, giving

$$m_x = \frac{Ei}{(1 - v^2)}\kappa_x$$

A further simplification is made by equating the term below the line to unity. This can be justified by the fact that Poisson's ratio is small. Further, if this approximation is applied to both m_x and m_y, they are both affected by the same amount. As it is the relative values of stiffness that affect the calculated bending moments and shear forces, such an adjustment has very little effect on the final results. The moment/curvature relationship then becomes

$$m_x = Ei\kappa_x \tag{5.48}$$

To achieve similitude of moments between a slab and the corresponding grillage, the stiffness terms of Equations 5.47 and 5.48 must be equated. This can clearly be achieved by adopting the same elastic modulus and second moment of area per unit breadth in the grillage as that of the slab.

A grillage member in torsion behaves according to the following well-known equation:

$$\phi = \frac{Tl}{GJ} \tag{5.49}$$

where ϕ is the angle of twist, T is the torque, l is the length of the beam, G is the shear modulus and J is the torsion constant (St. Venant constant). Figure 5.15 shows a portion of

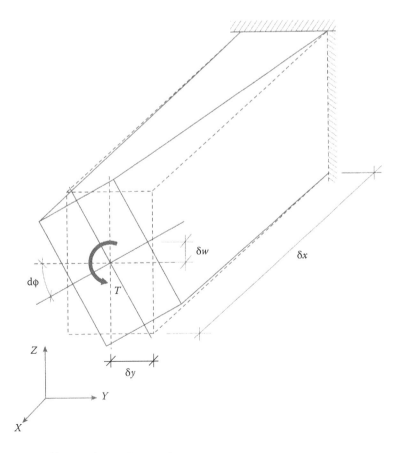

Figure 5.15 Segment of beam subjected to torsion.

a beam of length δx in torsion. The displacement in the Z direction is given by w and the angle of twist over the length δx is given by

$$d\phi = \frac{\partial w}{\partial y}$$

Hence,

$$\frac{d\phi}{dx} = \frac{\partial^2 w}{\partial x \partial y} \tag{5.50}$$

Substituting Equation 5.18 into Equation 5.50 gives

$$\frac{d\phi}{dx} = \kappa_{xy} \tag{5.51}$$

Applying Equation 5.49 to the beam of Figure 5.15 gives

$$d\phi = \frac{Tdx}{GJ}$$

$$\Rightarrow \quad T = GJ \frac{d\phi}{dx} \tag{5.52}$$

Substituting Equation 5.51 into Equation 5.52 gives

$$T = GJ\kappa_{xy}$$

This can be rewritten in terms of torque per unit breadth, t:

$$t = Gj^{\text{gril}}\kappa_{xy} \tag{5.53}$$

where j^{gril} is the torsion constant per unit breadth in the grillage member. Equation 5.42 gives an expression for the twisting moment per unit breadth in the bridge slab:

$$m_{xy} = G_{xy}j^{\text{slab}}\kappa_{xy} \tag{5.54}$$

To achieve similitude of twisting moments, m_{xy}, in the slab and torques, t, in the grillage members, the stiffness terms of Equations 5.53 and 5.54 are equated. This can be achieved by adopting the same shear modulus and torsion constant in the grillage member as is in the slab.

Equation 5.43 gives an expression for the torsion constant of the slab. Equating this to j^{gril} gives

$$j^{\text{gril}} = \frac{d^3}{6} \tag{5.55}$$

where d is the slab depth. Equation 5.55 ensures that the grillage members in both directions will have the same torsional constant per unit breadth. However, they will not necessarily have the same total torsional constant, as they may represent different breadths of slab if the grillage member spacings in the longitudinal and transverse directions differ. The torsion constant for the grillage member can alternatively be expressed in terms of the slab second moment of area:

$$j^{gril} = 2i^{slab} \tag{5.56}$$

Although Equations 5.55 and 5.56 are based on the grillage member having the same shear modulus as the slab, it will not generally be necessary to specify G_{xy} for the grillage model. The behaviour of a grillage member is essentially one-dimensional, and consequently, its shear modulus can be derived from the elastic modulus and Poisson's ratio directly using the well-known relationship

$$G_{xy} = \frac{E}{2(1+\nu)} \tag{5.57}$$

Typically, this is carried out automatically by the grillage program. The preceding derivation of grillage member torsional properties is applicable to thin plates of rectangular cross section where Equation 5.43 for the torsional constant is valid. Torsion in beams is complicated by torsional warping (in all but tubular sections), and formulas have been developed to determine an equivalent torsional constant for non-rectangular sections such that Equation 5.49 can be applied.

For rectangular beams with depth d and breadth b, the torsion constant is well known (e.g. Ghali et al. 2009):

$$J = bd^3 \left[\frac{1}{3} - 0.21\frac{d}{b}\left(1 - \frac{d^4}{12b^4}\right) \right]$$

When the breadth is greater than $10d$, the torsional constant per unit breadth may be approximated with

$$j \approx \frac{d^3}{3} = 4i \tag{5.58}$$

It can be seen that Equation 5.58 predicts a torsion constant for the beam which is twice that predicted by Equation 5.55 or 5.56 for isotropic slabs. The reason for this lies in the definition of torsion in a beam and of moment m_{xy} in a slab. Figure 5.16 shows a portion of a beam of breadth b and depth d in torsion. The shear stresses set up in the beam are shown in both horizontal and vertical directions. The torque in the beam results from both of these shear stresses and is given by

$$T = -\int_{-d/2}^{d/2} \tau_{xy}z \, dz - \int_{-b/2}^{b/2} \tau_{xz}y \, dy$$

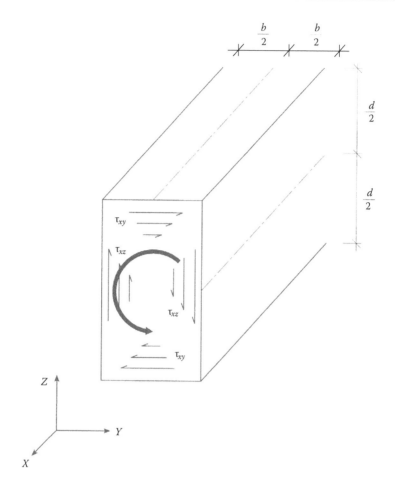

Figure 5.16 Beam subjected to torsion showing resulting shear stresses.

In the slab, Equation 5.36 shows that the moment m_{xy} is arrived at by summing only the shear stresses in the horizontal direction (i.e. only τ_{xy}). Consequently, the torsion constant for a grillage member representing a portion of an isotropic slab is only half that of a regular beam (or a grillage member representing a regular beam). In the slab, the shear stresses in the vertical direction are accounted for by the shear forces per unit breadth, q_x, as illustrated in Figure 5.17. It has been recommended that the edge grillage members be placed at 0.3 times the slab depth from the edge to coincide with the resultant of the shear stresses. The vertical shear stresses are accounted for in the grillage in the same manner by the shear forces q_y in the transverse beams.

5.3.3 Grillage member properties: Geometrically orthotropic slabs

Equation 5.34, reproduced here, applies to *materially* orthotropic slabs:

$$m_x = \frac{E_x i}{(1 - \nu_x \nu_y)} (\kappa_x + \nu_y \kappa_y)$$

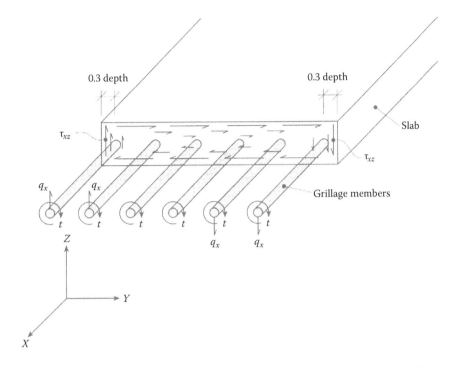

Figure 5.17 Slab with vertical shear stresses and corresponding grillage members with shear forces per unit breadth.

However, many bridges have the same modulus of elasticity, E, for both directions but are *geometrically* orthotropic, that is, they have different second moments of area per unit breadth in the orthogonal directions. It is common practice to use the equations developed for materially orthotropic thin plates to represent geometrically orthotropic bridges. In a grillage, this is achieved by basing the second moment of area per unit breadth in the X direction on the deck in that direction. Similarly, in the Y direction, the second moments of area per unit breadth for the grillage and the deck are equated.

Equation 5.40 stated that the two twisting moments at a point in a materially orthotropic slab are equal. Further, as stated in Equation 5.41, the two twisting curvatures are the same. It is assumed that the same conditions hold for geometrically orthotropic slabs, that is,

$$m_{xy} = m_{yx}$$

and

$$\kappa_{xy} = \kappa_{yx}$$

There is no facility in a grillage model to ensure that the two curvatures at a point are equal. However, in a fine grillage mesh, curvatures in the orthogonal directions at a point will be approximately equal. Then, if the same shear modulus and torsional constant are used in the two directions, it follows from Equation 5.42, reproduced and adapted here as Equation 5.59, that the twisting moments are equal:

$$\left. \begin{array}{l} m_{xy} = G_{xy}j/\kappa_{xy} \\ m_{yx} = G_{xy}j/\kappa_{yx} \end{array} \right\} \tag{5.59}$$

Hambly (1991) recommends using such a single torsional constant for both orthogonal directions:

$$j^{\text{gril}} = 2\sqrt{i_x^{\text{slab}} i_y^{\text{slab}}} \tag{5.60}$$

It can be seen that this equation is consistent with Equation 5.56 for an isotropic slab. The shear modulus for a slab made from one material, G, is a function of the elastic modulus E and Poisson's ratio ν. It is generally calculated internally in computer programs using Equation 5.57.

5.3.4 Computer implementation of grillages

There are many computer programs commercially available that are capable of analysing grillages. These programs are generally based on the same theory, the displacement or stiffness method, with some variations from program to program. The computer implementation of a plane grillage consists of defining a mesh of interconnected beams lying in one plane. The points at which these beams are connected are referred to as nodes. Each node has the capability to deflect vertically out-of-plane or to rotate about each axis of the plane. There is no facility for the nodes to deflect in either of the in-plane directions or to rotate about an axis perpendicular to the plane. The nodes are therefore said to have three degrees of freedom, two rotations and one translation. Consequently, in-plane axial forces are not modelled by the grillage. This inhibits the calculation of in-plane effects such as axial thermal expansion or contraction or in-plane prestressing. Such effects are normally determined separately and added to results from the grillage, according to the principle of superposition.

Some grillage programs allow, or require, the definition of a cross-sectional area for the beams. This may be used to define the bridge self-weight. In such cases, care should be taken to ensure that the self-weight is not applied twice by applying it to both the longitudinal and transverse beams. Some programs also use the cross-sectional area definition to model shear deformation. Even though the thin plate behaviour considered in Section 5.2 assumed that there was no shear deformation, some grillage programs do allow for shear deformation. This is generally achieved by defining a cross-sectional area and a shear factor, the product of which gives the shear area. While shear deformation is generally not very significant in typical bridges, it should improve the accuracy of the results if it is allowed for in the computer model. Some programs that allow the modelling of shear deformation will only give results of shear stresses when this option is invoked.

Grillage programs model the supports to the bridge slab as restraints at various nodes. It therefore makes sense, when formulating the grillage, to locate nodes at the centres of the bearings or supports. These nodal supports may be rigid, allowing no displacement or rotation in either of the two directions, or may allow one or more of these degrees of freedom. Most grillage programs will allow the use of spring supports and the imposition of specific support settlements. These facilities may be used to model the soil/structure interaction as discussed in Chapter 4.

5.3.5 Sources of inaccuracy in grillage models

It should always be borne in mind that the grillage analogy is only an approximation of the real bridge slab. Where the grillage is formulated without regard to the nature of the bridge slab, this approximation may be quite inaccurate, but when used correctly, it can simulate the behaviour reasonably well. However, even if due care is taken, some inherent inaccuracies exist in the grillage, a number of which are described here. It has been pointed out that the moments in two longitudinal or two transverse grillage members meeting end to end at a node will not necessarily be equal. The discontinuity between moments will be balanced by a discontinuity of torques in the beams in the other direction to preserve moment equilibrium at the node. Where only three beams meet at a node, such as where two longitudinal beams along the edge of a grillage meet one transverse beam, this discontinuity will be exaggerated. This is illustrated in Figure 5.18 where it can be seen that the torque T in the transverse beam, having no other transverse beam to balance it, corresponds to the discontinuity between the moments M_{b1} and M_{b2} in the longitudinal beams.

The required moment is arrived at by averaging the moments on either side of the node. The magnitude of these discontinuities can be reduced by choosing a finer grillage mesh. The same phenomenon causes discontinuities in torques and shears, which should be treated in the same way. As was mentioned earlier, excessively large discontinuities in moments, torques or shears indicate a grillage mesh which is too coarse, and requires the addition of more beams. The opposite of this is not necessarily true as other factors may also have an effect. Equation 5.34 gives an expression for moment per unit breadth, m_x, in the slab. This expression involves terms accounting for the curvature in the X and Y directions. When deriving the properties of a grillage member parallel to the X axis, the effect of curvature in the Y direction is ignored (see Equation 5.48). A similar simplification was made for m_y. As a result of this, the curvatures in the grillage members in one direction do not affect the moments in the beams in the other direction, in the same manner as they do in the actual bridge slab. This potential inconsistency is reduced by the low Poisson's ratio of bridge slab materials, which limits the influence of curvatures in one direction on moments in the orthogonal direction.

Equation 5.40 states that the moments m_{xy} and m_{yx} are equal in a slab as are the corresponding curvatures in the two directions. There is no mathematical or physical principle in the grillage to make this so. Torsions per unit breadth of similar magnitude in both directions in a grillage can be promoted by choosing the same torsional constant per unit breadth for the longitudinal and transverse beams. However, significant differences can remain.

Equations 5.45 and 5.46 provide expressions for the shear forces per unit breadth, q_x and q_y. The first of these equates the shear force per unit breadth q_x to the sum of two derivatives:

$$q_x = \frac{\partial m_x}{\partial x} + \frac{\partial m_{yx}}{\partial y}$$

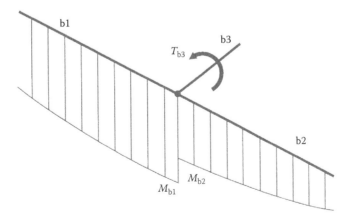

Figure 5.18 Distribution of bending moment in a segment of grillage mesh showing discontinuity in moment $(T_{b3} = M_{b1} - M_{b2})$.

In the grillage, the shear force in a longitudinal or transverse beam will simply be the derivative of the moment in that beam with respect to X or Y, whichever direction the beam lies in. There is no account taken of the derivative of the twisting moments, m_{xy} or m_{yx}. Fortunately, except for bridges with high skew, the magnitude of these moments is generally small.

5.3.6 Shear force near point supports

There is a particular problem in using grillage models to determine the intensity of shear force (shear force per unit breadth) near a discrete bearing. When bridges are supported at discrete intervals, there are sharp concentrations of shear intensity near each support. Each grillage member represents a strip of slab with the result that a point support at a node in a grillage model has an effective finite breadth. It follows that, if the grillage mesh density increases, the effective breadth decreases, and the calculated concentration of shear adjacent to the support increases. This direct relationship between mesh density and the calculated maximum shear intensity means that, if reasonably accurate results were to be obtained, the grillage member spacing would have to be selected near the support at the value that gives the correct result. In effect, the result is dictated by the member spacing; thus grillage analysis gives little useful information on shear intensity near point supports.

OBrien (1997) found that the grillage member spacing had a much reduced influence on the results for shear at distances of greater than or equal to a deck depth from the support. If it were assumed that shear enhancement was sufficient to cater for local concentrations of shear near a support, then grillage member spacing would have a much reduced importance. Thus, the designer would design for the shear force calculated at a deck depth from the support. Greater shear forces at points closer to the support would be ignored on the basis that load would be carried by direct compression rather than shear mechanisms.

5.3.7 Recommendations for grillage modelling

It is difficult to make specific recommendations on the use of a technique, such as grillage modelling, which is applicable to such a wide variety of structural forms. However, some general recommendations are valid for most grillage models. These should not be viewed as absolute and should be used in the context of good engineering judgement. Some more specific recommendations, such as those relating to voided or skewed bridge decks, are given in Chapter 6.

1. Longitudinal grillage members should be provided along lines of strength in the bridge slab, where these exist. Lines of strength may consist of concentrations of reinforcement, locations of prestressing tendons or beams in beam-and-slab bridges.
2. Where possible, grillage members should be located such that nodes coincide with the locations of supports to the bridge slab. The procedure of moving nodes locally to coincide with supports, illustrated in Figure 5.19a, should be avoided if possible, as this may result in skewed members, which complicate the interpretation of results. A variable mesh spacing (Figure 5.19b) is generally preferable.
3. There is little point in having longitudinal beams too closely spaced. Spacing will often be dictated by the location of supports or lines of strength in the bridge slab. A reasonable spacing of longitudinal beams is between one and two times the slab depth, and greater spacings are often possible without significant loss of accuracy.

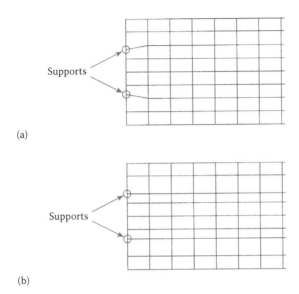

(a)

(b)

Figure 5.19 Alternative grillage meshes near point supports: (a) local adjustment to mesh near supports to maintain constant spacing of members elsewhere; (b) non-constant mesh spacing.

4. Transverse beams should have a spacing which is similar to that of the longitudinal beams. Often, this spacing will be greater than that of the longitudinal beams as the magnitude of moment in the transverse beams is generally relatively small. A choice of between one and three times the longitudinal spacing would be reasonable. The transverse grillage members should also be chosen to coincide with lines of transverse strength in the bridge slab, should they exist, such as heavily reinforced diaphragms above bridge piers.

5. If the spacing of grillage members is in doubt, a check can be performed by comparing the output of a grillage with that from a more refined grillage, that is, one with more longitudinal and transverse beams at a closer spacing. For bending moment results, increasing the mesh density tends (up to a point) to increase the accuracy.

6. It has been recommended by Hambly (1991) that the row of longitudinal beams at each edge of the grillage should be located in a distance of $0.3d$ from the edge of the slab, where d is the slab depth. The objective is to locate these beams close to the resultant of the vertical shear stresses, τ_{xz}, in the bridge slab, as illustrated in Figure 5.17. It has also been recommended that, when determining the torsional constant of these longitudinal grillage members, the breadth of slab outside $0.3d$ should be ignored. The second moments of area of these beams are calculated using the full breadth of slab in the normal way. The validity of this recommendation has been confirmed by the authors through comparisons of grillage analysis results with those of elaborate three-dimensional 'brick' FE models. Care should be taken, however, that this recommendation does not result in supports being placed in the wrong locations. Figure 5.20 illustrates an example where a member is correctly placed more than $0.3d$ from the end, so that the span length between supports in the grillage and the bridge slab are the same.

7. Supports to the grillage should be chosen to closely resemble those of the bridge slab. This may involve, for example, the use of elastic springs to simulate deformable bearings or ground conditions as discussed in Chapter 4.

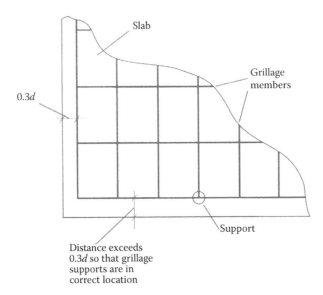

Figure 5.20 Segment of grillage mesh showing longitudinal members 0.3*d* from the edge, except for the end transverse members.

8. Beyond a deck depth from the face of the support, reasonable accuracy of shear forces per metre can be achieved with most sensible member spacings. Closer to the support, where shear enhancement occurs, grillage analysis is much less reliable.

5.4 PLANAR FINITE ELEMENT ANALYSIS OF SLAB DECKS

The FE method was pioneered in the mid-1950s for use mainly in the aeronautical industry. Originally, it was used for in-plane analysis of structures, but it was soon extended to the problem of plate bending by Zienkiewicz and Cheung (1964). Much development has taken place since this pioneering work, and many texts now exist that give a comprehensive description of the method (see, for example, that of Zienkiewicz et al. 2013). FE analysis is relatively easy to use and comprehend and, when applied correctly, is at least as accurate as, and generally more accurate than, the grillage method.

Ghali et al. (2009) give a concise description of FE theory as applied to problems in structural analysis. FE uses the displacement or stiffness method of analysis, with the element stiffnesses being combined to give the global stiffness matrix of the entire structure. The elements are joined at nodes, and the stiffness method should ensure compatibility of displacements between elements at each node.

5.4.1 FE theory: Beam elements

Element behaviour derives from its displacement function, an assumed shape that it takes up in response to the displacements at its nodes. Perhaps the simplest example of an element is the beam, and the simplest displacement function is the cubic polynomial. Referring to Figure 5.21, the beam is assumed to deform according to the cubic polynomial:

$$w = A_1 + A_2 x + A_3 x^2 + A_4 x^3$$

Figure 5.21 Beam element.

In matrix form, this is written as

$$w = [1,\ x,\ x^2,\ x^3] \begin{Bmatrix} A_1 \\ A_2 \\ A_3 \\ A_4 \end{Bmatrix}$$

or

$$w = [P]\{A\} \tag{5.61}$$

where

$$[P] = [1,\ x,\ x^2,\ x^3] \text{ and } \{A\} = \begin{Bmatrix} A_1 \\ A_2 \\ A_3 \\ A_4 \end{Bmatrix}$$

The rotations are found by differentiating the translations. Hence,

$$\theta = [0,\ 1,\ 2x,\ 3x^2] \begin{Bmatrix} A_1 \\ A_2 \\ A_3 \\ A_4 \end{Bmatrix}$$

The degrees of freedom or nodal displacements of this element are the translations and rotations at each end:

$$\{D^*\} = \begin{Bmatrix} w_{x=0} \\ \theta_{x=0} \\ w_{x=l} \\ \theta_{x=l} \end{Bmatrix} = \begin{bmatrix} 1 & 0 & 0 & 0 \\ 0 & 1 & 0 & 0 \\ 1 & l & l^2 & l^3 \\ 0 & 1 & 2l & 3l^2 \end{bmatrix} \begin{Bmatrix} A_1 \\ A_2 \\ A_3 \\ A_4 \end{Bmatrix}$$

or

$$\{D^*\} = [C]\{A\} \tag{5.62}$$

where

$$[C] = \begin{bmatrix} 1 & 0 & 0 & 0 \\ 0 & 1 & 0 & 0 \\ 1 & l & l^2 & l^3 \\ 0 & 1 & 2l & 3l^2 \end{bmatrix}$$

The nodal displacements are taken to be known and are used to find the coefficients of the polynomial:

$$\{A\} = [C]^{-1}\{D^*\} \tag{5.63}$$

The displacement function can be expressed in terms of the nodal displacements by substituting Equation 5.63 into 5.61:

$$w = [P][C]^{-1}\{D^*\} \tag{5.64}$$

For a beam, the strain distribution is described by the curvature. Hence, the generalised strain vector is simply the curvature, given by

$$\{\varepsilon\} = \{\kappa\} = -\left\{\frac{d^2 w}{dx^2}\right\}$$

This can be found by differentiating the displacement function of Equation 5.64. As [C] and $\{D^*\}$ are matrices of constants, only the coefficients of [P] need to be differentiated:

$$\{\varepsilon\} = [0, \quad 0, \quad -2, \quad -6x][C]^{-1}\{D^*\} \tag{5.65}$$

In FE theory, Equation 5.65 is often expressed in the form

$$\{\varepsilon\} = [B]\{D^*\}$$

where

$$[B] = [0, \quad 0, \quad -2, \quad -6x][C]^{-1}$$

For a beam, the stress distribution is described by the bending moment. Thus, the generalised stress vector in this case is simply the moment:

$$\{\sigma\} = \{M\} = EI\{\kappa\}$$

where EI is the product of Young's modulus and second moment of area. In FE theory, the stress/strain equation is written as

$$\{\sigma\} = [d]\{\varepsilon\}$$

where, in this case, [d] is a 1×1 matrix containing the product, EI.

Ghali et al. (2009) show that the element-stiffness matrix can be found by minimising the total potential energy and is given by

$$[K] = \int_0^l [B]^T[d][B]dx$$

Substituting for [B] and [d] and integrating gives the stiffness matrix for a beam in bending:

$$[K] = EI \begin{bmatrix} \dfrac{12}{l^3} & \dfrac{6}{l^2} & -\dfrac{12}{l^3} & \dfrac{6}{l^2} \\[2mm] \dfrac{6}{l^2} & \dfrac{4}{l} & -\dfrac{6}{l^2} & \dfrac{2}{l} \\[2mm] -\dfrac{12}{l^3} & -\dfrac{6}{l^2} & \dfrac{12}{l^3} & -\dfrac{6}{l^2} \\[2mm] \dfrac{6}{l^2} & \dfrac{2}{l} & -\dfrac{6}{l^2} & \dfrac{4}{l} \end{bmatrix}$$

For a structure made up of beams, FE analysis is, in effect, identical to an analysis using the stiffness method. For slab bridges, FE theory allows the element to be two-dimensional. The bridge slab is modelled as a finite number of discrete slab elements, sometimes complemented with beam elements, joined at a finite number of nodes. In the simpler models, all of the elements lie in one plane.

5.4.2 FE theory: Plate elements

For bridge slabs, the most common types of element used are rectangular or parallelogram in shape, although triangular elements are sometimes also necessary. A four-node rectangular element is described here as it demonstrates the main features of this kind of element. Three degrees of freedom are allowed at each node: vertical translation and rotation about two axes (Figure 5.22).

As for the beam element, everything starts with the displacement function, and, in many ways, the choice of a displacement function determines whether or not the element

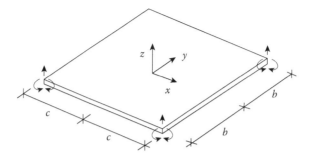

Figure 5.22 Plate FE.

accurately represents true slab behaviour. With four nodes and three degrees of freedom per node, there are 12 nodal displacements in this case; thus there are 12 components in [D*]. It follows that [C] must be a 12 × 12 matrix, and there can only be 12 components in {A}. A complete fourth-order polynomial in two variables has 15 terms; thus not all terms can be used. Ghali et al. (2009) propose the following compromise:

$$w = A_1 + A_2x + A_3y + A_4x^2 + A_5xy + A_6y^2$$
$$+ A_7x^3 + A_8x^2y + A_9xy^2 + A_{10}y^3 + A_{11}x^3y + A_{12}xy^3 \tag{5.66}$$

that is, $w = [P]\{A\}$, where

$$[P] = [1 \ \ x \ \ y \ \ x^2 \ \ xy \ \ y^2 \ \ x^3 \ \ x^2y \ \ xy^2 \ \ y^3 \ \ x^3y \ \ xy^3] \tag{5.67}$$

Along the sides of the rectangular element, this function degenerates into a beam polynomial. For example, at $x = c$ in Figure 5.22, it becomes

$$w = A_1 + A_2c + A_3y + A_4c^2 + A_5cy + A_6y^2 + A_7c^3 + A_8c^2y$$
$$+ A_9cy^2 + A_{10}y^3 + A_{11}c^3y + A_{12}cy^3$$
$$= (A_1 + A_2c + A_4c^2 + A_7c^3) + (A_3 + A_5c + A_8c^2 + A_{11}c^3)y$$
$$+ (A_6 + A_9c)y^2 + (A_{10} + A_{12}c)y^3$$

that is, a cubic polynomial, as assumed for the beam element. This is useful as a cubic polynomial can be uniquely defined by the translations and rotations at each end (four equations in four unknowns). Hence, adjacent elements can have identical shapes along their boundary and therefore identical curvatures and bending moments. This is illustrated in Figure 5.23 – the shape at the boundary that joins nodes 1 and 2 is identical.

The same is true on the boundary at $x = -c$, and on the boundaries at $y = \pm b$ – Figure 5.23 shows identical shapes of the boundary between nodes 1 and 4 in adjacent elements. Unfortunately, there are not enough equations to ensure these continuities while also ensuring continuity of

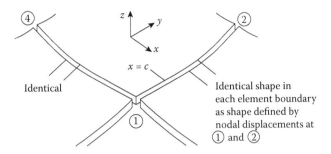

Figure 5.23 Continuity at nodes of plate elements.

slopes normal to the boundary. Equation 5.66 can be differentiated to find the slope with respect to x:

$$\frac{\partial w}{\partial x} = A_2 + 2A_4x + A_5y + 3A_7x^2 + 2A_8xy + A_9y^2 + 3A_{11}x^2y + A_{12}y^3$$

Along the edge at $x = c$, this becomes

$$\left.\frac{\partial w}{\partial x}\right|_{(x=c)} = A_2 + 2A_4c + A_5y + 3A_7c^2 + 2A_8cy + A_9y^2 + 3A_{11}c^2y + A_{12}y^3$$

$$= (A_2 + 2A_4c + 3A_7c^2) + (A_5 + 2A_8c + 3A_{11}c^2) + A_9y^2 + A_{12}y^3$$

that is, a cubic polynomial. There is a value for this normal slope at each end of the boundary that has not already been used, but these two new values are not enough to define the function uniquely. It follows that full compatibility of all slopes cannot be guaranteed between adjacent elements. This is illustrated in Figure 5.23 as a possible discontinuity in slope along the boundary between nodes 1 and 2. The same applies at $x = -c$ and for the other normal slope, $\partial w/\partial y$, at $y = \pm b$. Despite this problem, the 12-node plate element has been shown to give generally good results and is widely used.

As for the beam element, the rotations are found by differentiating the translations. Substituting for x and y at the nodes gives an equation of the form

$$\{D^*\} = [C]\{A\} \qquad (5.68)$$

However, $\{D^*\}$ now has 12 components, one translation and two rotations at each node. As before, Equation 5.68 is used to find the coefficients of the displacement function:

$$\{A\} = [C]^{-1}\{D^*\} \qquad (5.69)$$

and hence,

$$w = [P][C]^{-1}\{D^*\} \tag{5.70}$$

As for the beam element, the strain distributions are described by curvatures, but there are now three curvatures. The generalised strain vector for a plate element is given by

$$\{\varepsilon\} = \begin{Bmatrix} \kappa_x \\ \kappa_y \\ \kappa_{xy} \end{Bmatrix} = \begin{Bmatrix} -\dfrac{\partial^2 w}{\partial x^2} \\ -\dfrac{\partial^2 w}{\partial y^2} \\ \dfrac{\partial^2 w}{\partial x \partial y} \end{Bmatrix} \tag{5.71}$$

As before, these are found by differentiating the displacement function. Differentiating the components of [P] in Equations 5.67 and 5.70 gives

$$\{\varepsilon\} = [B]\{D^*\}$$

where

$$[B] = \begin{bmatrix} 0 & 0 & 0 & -2 & 0 & 0 & -6x & -2y & 0 & 0 & -6xy & 0 \\ 0 & 0 & 0 & 0 & 0 & -2 & 0 & 0 & -2x & -6y & 0 & -6xy \\ 0 & 0 & 0 & 0 & 1 & 0 & 0 & 2x & 2y & 0 & 3x^2 & 3x^2 \end{bmatrix} [C]^{-1} \tag{5.72}$$

For orthotropic plates, the generalised stress vector consists of the moments per metre. The moment curvature relationships are taken from Equations 5.34, 5.35 and 5.42, and are presented here in matrix form:

$$\{\sigma\} = \begin{Bmatrix} m_x \\ m_y \\ m_{xy} \end{Bmatrix} = \begin{bmatrix} \dfrac{E_x i}{1-v_x v_y} & \dfrac{v_y E_x i}{1-v_x v_y} & 0 \\ \dfrac{v_x E_y i}{1-v_x v_y} & \dfrac{E_y i}{1-v_x v_y} & 0 \\ 0 & 0 & G_{xy} j \end{bmatrix} \begin{Bmatrix} \kappa_x \\ \kappa_y \\ \kappa_{xy} \end{Bmatrix}$$

that is,

$$\{\sigma\} = [d]\{\varepsilon\}$$

where

$$[d] = \begin{bmatrix} \dfrac{E_x i}{1-v_x v_y} & \dfrac{v_y E_x i}{1-v_x v_y} & 0 \\[3mm] \dfrac{v_x E_y i}{1-v_x v_y} & \dfrac{E_y i}{1-v_x v_y} & 0 \\[3mm] 0 & 0 & G_{xy} j \end{bmatrix} \tag{5.73}$$

As for beams, the element stiffness matrix is found by minimising the potential energy, but, in this case, integration is over two dimensions:

$$[K] = \iint [B]^T [d][B] dx\, dy$$

where [B] and [d] are given by Equations 5.72 and 5.73, respectively.

5.4.3 Similitude between plate FE model and bridge slab

The moments per unit breadth, m_x, m_y and m_{xy}, are output directly by FE programs. These are calculated internally at the centres of the elements and are generally given at these points. Many programs provide the ability to determine moments at nodes or at other points using interpolation. If this facility is used, a check is useful to ensure that the values given are consistent with those at the neighbouring elements. Equations 5.34 and 5.35 give expressions for the moments, m_x and m_y, in a thin plate. Each of these expressions involves terms relating to the curvature in both the X and Y directions. The FEs will behave according to these equations and, unlike a grillage analysis, will account for the effect of curvature in one direction on the moment in the other. This is a significant advantage of the FE method over the grillage approach.

Equation 5.42 gives an expression for the moments, m_{xy} and m_{yx}, in a thin, materially orthotropic plate. The FEs will satisfy this equation, and the problem inherent in grillage modelling of torques per unit breadth not being equal in orthogonal directions does not arise.

Equations 5.45 and 5.46 give expressions for the shear force per unit breadth in a thin plate. These expressions involve derivatives of the direct moment m_x (or m_y) and the twisting moment m_{yx} (or m_{xy}). It was shown above that a grillage model does not take account of the derivative of the twisting moment. In FE analysis, shear force per unit breadth can be calculated, although not all programs offer this facility. The twisting moment term can readily be accounted for, although in some programs it may not be. Where the twisting moments are significant, it is advisable to determine whether or not shear forces are calculated correctly using Equations 5.45 and 5.46.

The element described in Section 5.4.2 does not model in-plane distortion, and consequently, each node has three degrees of freedom, out-of-plane translation and rotation about both in-plane axes. No particular problem arises from using elements which allow in-plane deformations in addition to out-of-plane bending, that is, five or six degrees of freedom per node. However, the model should not be under-restrained. The support arrangement must ensure that the structure is stable; it should prevent free body motion in either of the in-plane directions or rotation in that plane. At the same time, the model must not be over-restrained. It should allow the movements that will be allowed by sliding bearings.

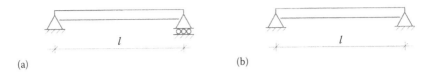

Figure 5.24 Simply supported beams: (a) pinned–rollers; (b) pinned–pinned.

Figure 5.24 illustrates the two-dimensional equivalent – if there is a sliding bearing at the right-hand support, the FE model should allow this in-plane movement (Figure 5.24a) as constraining it (Figure 5.24b) may give incorrect results. FE models that allow in-plane distortions are generally only necessary if in-plane distortion is important. This might arise in a bridge curved in plan, when analysed for the effects of restrained axial temperature or prestress.

FE models in which the elements are not all located in one plane can be used to model bridge decks which exhibit significant three-dimensional behaviour. Some of these types of model are discussed in Chapter 7.

5.4.4 Properties of plate finite elements

The material properties of the elements are defined in relation to the material properties of the bridge slab. When materially orthotropic FEs are used, five elastic constants, E_x, E_y, G_{xy}, ν_x and ν_y, typically need to be specified. Some programs assume a value for G_{xy} based on the values input for the other four elastic constants. If this is the case, the validity of this relationship should be checked for the particular plate under consideration.

Isotropic bridge slabs

In the case of bridges which are idealised as isotropic plates, only two elastic constants need to be defined for the FEs, E and ν. The shear modulus, G, is determined by the program from these constants directly according to Equation 5.57. As the element is of constant depth, the second moment of area per unit breadth is given by Equation 5.33:

$$i = \frac{d^3}{12}$$

In a typical program, the user simply specifies the element depth as

$$d = \sqrt[3]{12i} \tag{5.74}$$

Geometrically orthotropic bridge slabs

Geometrically orthotropic bridge decks are frequently modelled using materially orthotropic FEs. In such cases, $i_x \neq i_y$, but only one depth can be specified. This problem can be overcome by determining an equivalent plate depth and altering the moduli of elasticity of the element to allow for the differences in the second moment of area.

Equation 5.34 gives an expression for the moment, m_x, in a materially orthotropic FE:

$$m_x = \frac{E_x^{\text{elem}} i^{\text{elem}}}{(1 - v_x v_y)} (\kappa_x + v_y \kappa_y)$$

where E_x^{elem} and i^{elem} are the element elastic modulus and second moment of area per unit breadth, respectively. Equation 5.35 gives a similar expression for m_y. In most geometrically orthotropic bridge slabs, there is only one modulus of elasticity, E_{slab}, for both directions, but there are two second moments of area per unit breadth, i_x^{slab} and i_y^{slab}. However, similitude between the FE and the bridge slab can be achieved by keeping the products of elastic modulus and second moment of area equal:

$$E_x^{\text{elem}} i^{\text{elem}} = E^{\text{slab}} i_x^{\text{slab}} \tag{5.75}$$

$$E_y^{\text{elem}} i^{\text{elem}} = E^{\text{slab}} i_y^{\text{slab}} \tag{5.76}$$

The modulus of elasticity of the element in the X direction may be chosen arbitrarily to be equal to the modulus of elasticity of the bridge slab, that is,

$$E_x^{\text{elem}} = E^{\text{slab}}$$

Substituting this into Equations 5.75 and 5.76 gives

$$i^{\text{elem}} = i_x^{\text{slab}} \tag{5.77}$$

and

$$E_y^{\text{elem}} = E^{\text{slab}} \frac{i_y^{\text{slab}}}{i_x^{\text{slab}}} \tag{5.78}$$

The equivalent element depth can be calculated from Equation 5.74. For a materially orthotropic slab, the moment/curvature relationship for the twisting moment, m_{xy}, is given by Equation 5.42. An approximate expression for the constant, G_{xy}, has been suggested by Troitsky (1967):

$$G_{xy}^{\text{elem}} = \frac{\left(1 - \sqrt{v_x v_y}\right) \sqrt{E_x^{\text{elem}} E_y^{\text{elem}}}}{2(1 - v_x v_y)} \tag{5.79}$$

For a geometrically orthotropic slab with a single modulus of elasticity and Poisson's ratio, a similar expression can be determined by substituting from Equations 5.75 and 5.76 to give

$$G_{xy}^{\text{elem}} = \frac{E^{\text{slab}}}{2(1 + v)} \sqrt{\frac{i_x^{\text{slab}} i_y^{\text{slab}}}{(i^{\text{elem}})^2}} \tag{5.80}$$

To be consistent with the equations for E_x^{elem} and i^{elem} derived above, Equation 5.77 applies and Equation 5.80 becomes

$$G_{xy}^{elem} = \frac{E^{slab}}{2(1+\nu)}\sqrt{\frac{i_y^{slab}}{i_x^{slab}}} \qquad (5.81)$$

Equation 5.79 was derived by assuming an average value of the elastic moduli in the two directions and an average Poisson's ratio. Consequently, the accuracy of this and Equation 5.81 diminishes as the variation in the elastic properties in the two directions increases. In such cases, the shear modulus may need to be reduced. It was reported by Troitsky (1967), from the results of analysis and experimentation on steel orthotropic bridge decks, that the shear modulus given by the above expression may need to be reduced by a factor of between 0.5 and 0.3. The lower value of 0.3 was reported to come from an extreme case where the flexural stiffness in the two directions varied by a factor of 20.

To determine if the influence of the shear modulus on the analysis is significant, the authors would suggest analysing the orthotropic plate using a value predicted by Equation 5.81 and analysing again using a shear modulus of half this value. As an alternative, the orthotropic nature of the plate might be better handled using a combination of elements and beam members or a three-dimensional model. These types of model are discussed further in Chapters 6 and 7.

The expressions given above relate to bridge slabs with the same modulus of elasticity in both directions but can easily be modified where this is not the case.

5.4.5 Shear forces in plate FE models

For the modelling of bridge slabs, plate FE models are a good compromise between accuracy and complexity. However, while they give good results for bending moment, the results for shear near discrete supports are far less satisfactory. The problem is best illustrated with the example illustrated in Figure 5.25. This is a 14 m span solid slab with two point supports at each end. While in practice, the support provided by pot or spherical bearings is distributed over a small area, it is represented here by a point. A $0.5 \times 0.5 \times 0.1$ m³ solid block is placed under the main slab to allow some dispersion of the support over an assumed 0.5×0.5 m² bearing area.

An accurate analysis of this slab was carried out by Harney (2012) using a full three-dimensional FE model with solid 'brick' elements. For uniformly distributed loading of 24 kN/m², the shear intensities – shear forces per unit transverse width – are illustrated in Figure 5.25b. The peaks of shear intensity reach approximately 450 kN/m over each point support. Such an analysis gives accurate results but is impractical for everyday analysis and design. Analysis using simpler plate FE models gives the results illustrated in Figure 5.26. In this case, the calculated maximum shear intensities are unrealistically large and get larger as the mesh density increases. The peaks over the supports reach 1300 kN/m² for the mesh of Figure 5.26b and would be even greater if the FE mesh were made denser. The reason is that the plate FE model has no depth through which the point force reaction can disperse. With a force on an infinitely small area, the shear stresses in the immediate vicinity of the support tend towards infinity. Fortunately, this is a local effect, and the plate FE model gives good agreement with the more accurate three-dimensional FE model at one depth away from the supports. If it is assumed that shear enhancement is sufficient to deal with the shear intensities within one depth of the support, then a plate FE model appears to be sufficiently accurate to find the shear intensities at other points.

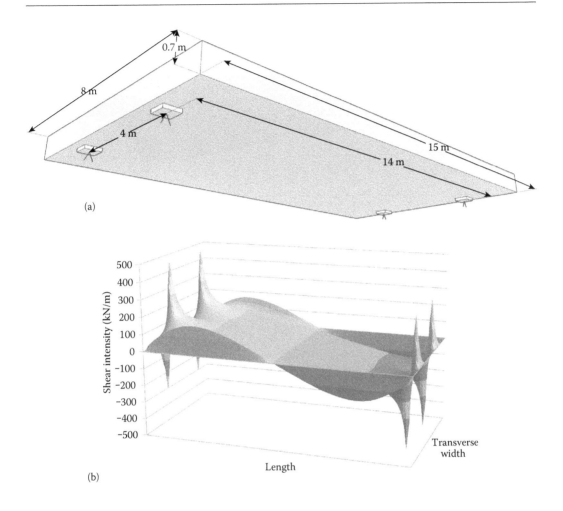

(a)

(b)

Figure 5.25 Solid bridge slab with point supports: (a) geometry; (b) shear forces per unit width, q_x, from 3-D solid FE analysis.

5.4.6 Recommendations for FE analysis

There are many commercially available computer programs for FE analysis of bridge decks. The implementation of the FE model is carried out in a similar manner to a grillage, and many of the comments in Section 5.3 apply. One variation between the two methods is that the FE model may allow for in-plane deformations, and consequently, the nodes will often have five or six degrees of freedom. This type of model is useful when the slab is not all in one plane.

As with grillage modelling, it is difficult to make specific recommendations relating to FE modelling of bridge slabs, but some general guidelines are given here. Once again, these should not be viewed as absolute. In general, more elements tend to result in greater accuracy, although this is not guaranteed. Many engineers use denser meshes of elements in

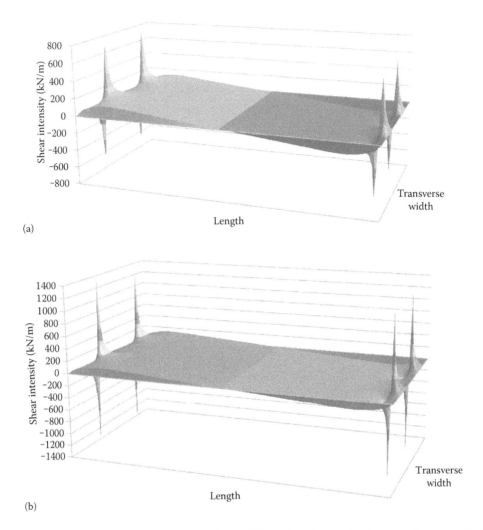

Figure 5.26 Plate FE analyses of the solid slab of Figure 5.25 with uniform grid of square elements: (a) with 0.125 × 0.125 m² elements; (b) with 0.0625 × 0.0625 m² elements.

those parts of a bridge where bending moment changes rapidly, such as near an interior support. While this makes sense, it is often more convenient if a consistent mesh density is used throughout a bridge. Provided it is dense enough at the critical locations, this is a sensible strategy.

Unlike the grillage method, the FE response to applied loading is based on an assumed displacement function. This function may be applicable to elements of a certain shape only, and the program may allow the user to define elements that do not conform to this shape. Considering, for example, quadrilateral elements with nodes at the four corners, a typical program may be able to deal with approximately square, rectangular or parallelogram elements, as shown in Figure 5.27a. On the other hand, programs may give an inaccurate representation for elements with poor length to breadth ratios or obtuse angles as shown in Figure 5.27b. More specific recommendations are given below, and further guidance, applicable to voided and skewed bridge decks, is given in Chapter 6.

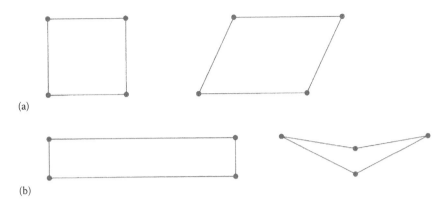

Figure 5.27 Possible shapes of quadrilateral FEs: (a) generally good shapes; (b) potentially problematic shapes.

1. Regularly shaped FEs should be used where possible. These should be as 'square' as possible in the case of quadrilateral elements and should tend towards equilateral triangles in the case of triangles. Obviously, considerable deviation from these shapes may be permissible, and the documentation provided with the program should be consulted for specific recommendations. In the absence of information to the contrary, two rules commonly applied to quadrilateral elements are that the ratios of the side lengths should not exceed approximately 2:1 and that no two sides should have an internal angle greater than approximately 135°. Hence, the elements of Figure 5.27b are not good shapes.

2. Mesh discontinuities should be avoided. These may occur when attempting to refine the mesh, such as in Figure 5.28a, where elements 1 and 2 are connected to each other at point P but are not connected to element 4. Some elements have mid-side nodes so that it is possible for example to have elements 3 and 4 connected to the mid-side node of element 1 at Q. A mesh is shown in Figure 5.28b where mid-side nodes are not needed, and all elements are connected.

3. The spacing of elements in the longitudinal and transverse directions should be similar. This should happen automatically if the first recommendation is adhered to.

4. There is little point in using too many elements as an excessive number slow the running of the program and may not result in significantly greater accuracy. If mesh density is in question, it is useful to compare the output of a model with the chosen mesh density to that of a model with a greater density. Similar results from the two suggest that the mesh is sufficiently dense.

5. Elements should be located so that nodes coincide with the bearing locations. This is generally easily achieved.

6. The mesh should ideally be designed so that nodes coincide with the points where loading is applied, and element centres coincide with the points where the results are required. This is what happens internally; thus not doing it means that some form of interpolation will be used in applying loading to nodes or in calculating results. However, these requirements are often contradictory, in which case, a fine mesh density will minimise the amount of interpolation required.

7. Supports to the FE model should be chosen to closely resemble those of the bridge slab. This may involve, for example, the use of elastic springs to simulate deformable bearings or ground conditions, as discussed in Chapter 4.

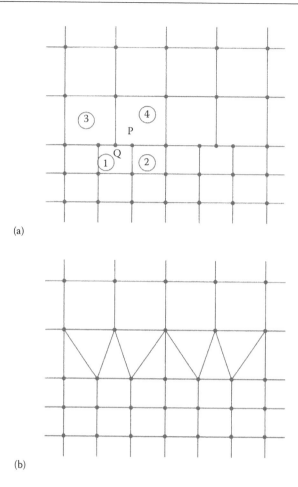

(a)

(b)

Figure 5.28 Meshes of FEs at transition between coarse and dense meshes: (a) potentially problematic arrangement; (b) good arrangement.

8. Shear forces near point supports in plate FE models tend to be unrealistically large and should be treated with scepticism. This problem will *not* be improved by using a finer mesh density. Fortunately, results at more than a deck depth away from the support have been found in many cases to be reasonably accurate (OBrien et al. 1997; Harney 2012).

5.5 WOOD AND ARMER EQUATIONS

Much of this chapter has been concerned with methods of analysis of slab bridges. The results of such analyses give three components of bending moment at each point, m_x, m_y and m_{xy}. This section addresses the design problem of how the engineer should calculate the moment capacity required to resist such moments. The so-called 'Wood and Armer' equations are credited to Wood (1968) and Armer (1968). In more recent years, Denton and Burgoyne (1996) extended the concept to include cases of bridge assessment, that is, cases where the moment capacities are known.

The derivation of the Wood and Armer equations follows from the fact that, as the bending moments, m_x, m_y and m_{xy}, are all vectors, they are combinable using vector addition in a manner similar to the concept of Mohr's circle of stresses. Resultant moments can be calculated at any angle of orientation and can, if excessive, result in yield of the slab at any such angle.

A small segment of slab is illustrated in Figure 5.29a, and the possibility is considered of failure on a face, AB, at an angle of θ to the Y axis. The length of the face AB is l, and as can be seen in the figure, the projected lengths on the X and Y axes are $l \sin\theta$ and $l \cos\theta$, respectively. The moment per unit length on the X face is m_x so the moment on BC is $m_x(l \cos\theta)$. The corresponding moment on AC is $m_y(l \sin\theta)$. These moments are illustrated in Figure 5.29b using double-headed arrows to denote bending moment, where the moment is about the axis of the arrow. The twisting moments per unit length, m_{xy} and m_{yx}, are also illustrated in this figure.

The vectors representing the moments are resolved to determine the moments on the face, AB. For convenience, a second axis system, N–T, is introduced where N is normal to the face AB and T is parallel (tangential) to it. The direct moment per unit length on AB is denoted m_n, and the twisting moment per unit length is denoted m_{nt}. All vectors are resolved parallel and perpendicular to AB in Figure 5.30.

Considering components parallel to AB first:

$$m_n l = m_x l \cos^2\theta + m_y l \sin^2\theta + m_{xy} l \cos\theta \sin\theta + m_{yx} l \sin\theta \cos\theta$$

$$\Rightarrow \quad m_n = m_x \cos^2\theta + m_y \sin^2\theta + 2m_{xy} \cos\theta \sin\theta \qquad (5.82)$$

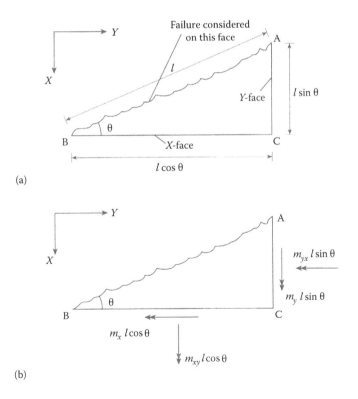

Figure 5.29 Plan view of segment of slab: (a) geometry; (b) applied axial and twisting moments.

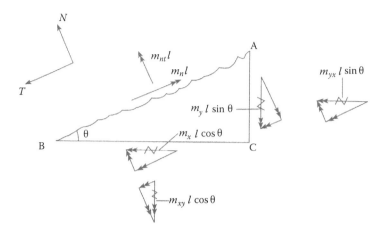

Figure 5.30 Resolution of moments on a segment of slab parallel and perpendicular to AB.

Considering components perpendicular to AB gives

$$m_{nt}l = m_y\, l \sin\theta \cos\theta - m_x\, l \cos\theta \sin\theta + m_{xy}\, l \cos^2\theta - m_{yx}\, l \sin^2\theta$$

$$\Rightarrow \quad m_{nt} = (m_y - m_x)\cos\theta \sin\theta + m_{xy}(\cos^2\theta - \sin^2\theta) \tag{5.83}$$

In an orthotropic steel plate, moment capacity is generally provided in two orthogonal directions. In a concrete slab, ordinary or prestressing reinforcement is provided in two directions, which are not necessarily orthogonal. In this section, only orthogonal systems of reinforcement are considered.

An orthogonal system of reinforcement provides moment capacity in two perpendicular directions, which are taken here to be parallel to the coordinate axes. Hence, the moment capacities per unit length can be expressed as m_x^* and m_y^* as illustrated in Figure 5.31. This figure is different from Figure 5.29 in that there are no twisting moment terms; no capacity to resist twisting moment is assumed to be provided. Equation 5.82 gives the moment on

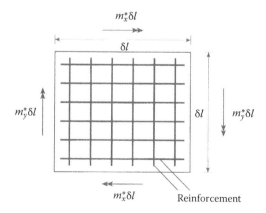

Figure 5.31 Segment of slab illustrating the moment capacities provided.

a face at an angle θ to the Y axis. A corresponding equation can readily be derived for the moment capacity. Leaving out the m_{xy} term in Equation 5.82 leads to

$$m_n^* = m_x^* \cos^2 \theta + m_y^* \sin^2 \theta \qquad (5.84)$$

To prevent failure on face AB of Figure 5.29, the moment capacity must exceed the applied moment, that is,

$$m_n^* \geq m_n$$

Substituting from Equations 5.82 and 5.84 gives

$$m_x^* \cos^2 \theta + m_y^* \sin^2 \theta \geq m_x \cos^2 \theta + m_y \sin^2 \theta + 2m_{xy} \cos\theta \sin\theta$$
$$\Rightarrow \quad f(\theta) \geq 0$$

where

$$f(\theta) = \left(m_x^* - m_x\right)\cos^2 \theta + \left(m_y^* - m_y\right)\sin^2 \theta - 2m_{xy}\cos\theta\sin\theta \qquad (5.85)$$

The function $f(\theta)$ is the excess moment capacity for the angle θ, that is, the amount by which the moment capacity exceeds the applied moment for that angle. To prevent failure of the slab, it is clearly necessary that this function exceeds zero for all values of θ. The most critical angle will be that for which $f(\theta)$ is a minimum. This minimum value is found by differentiating the function and equating to zero:

$$\frac{\mathrm{d}f(\theta)}{\mathrm{d}\theta} = 0$$

$$\Rightarrow \quad \hat{k}\left[\left(m_y^* - m_y\right) - \left(m_x^* - m_x\right)\right] + (\hat{k}^2 - 1)m_{xy} = 0 \qquad (5.86)^*$$

where $\hat{k}$ is a critical value for $k = \tan\theta$. For this to be a minimum excess moment capacity rather than a maximum, the second derivative of $f(\theta)$ must be positive:

$$\frac{\mathrm{d}^2 f(\theta)}{\mathrm{d}\theta^2} > 0$$

$$\Rightarrow \quad 2\hat{k}m_{xy} > (\hat{k}^2 - 1)\left[\left(m_y^* - m_y\right) - \left(m_x^* - m_x\right)\right]$$

---

* In the original paper, Wood (1968) divided Equation 5.85 across by $\cos^2\theta$ before differentiating, which resulted in a different expression for the critical angle.

EXAMPLE 5.1   MOMENT CAPACITY CHECK

At a point in a bridge slab, the moments per unit length due to applied loads have been found to be $m_x = 80$, $m_y = 190$ and $m_{xy} = 20$. It is required to determine if moment capacities of $m_x^* = 110$ and $m_y^* = 250$ are sufficient.

Equation 5.86 is used to determine the angle for which the excess moment capacities are critical:

$$\hat{k}\left[\left(m_y^* - m_y\right) - \left(m_x^* - m_x\right)\right] + (\hat{k}^2 - 1)m_{xy} = 0$$

$$\Rightarrow \quad (60 - 30)\hat{k} + 20(\hat{k}^2 - 1) = 0$$

$$\Rightarrow \quad 20\hat{k}^2 + 30\hat{k} - 20 = 0$$

The critical root of this equation is when m_{xy} and $\hat{k}$ are of the same sign. As m_{xy} is positive in this example, the positive root is the critical one: $\hat{k} = 0.5$. Hence, the critical angle is $\tan^{-1}(0.5) = 27°$. The minimum excess capacity is then found by substitution in Equation 5.85:

$$f(\theta) = \left(m_x^* - m_x\right)\cos^2\theta + \left(m_y^* - m_y\right)\sin^2\theta - 2m_{xy}\cos\theta\sin\theta$$
$$= 30\cos^2(27°) + 60\sin^2(27°) - 2 \times 20\cos(27°)\sin(27°)$$
$$= 20$$

As the excess capacity at the critical angle is positive, the slab has sufficient bending capacity to resist m_n for all values of θ.

Substituting back in from Equation 5.86, it can be shown that this inequality is satisfied if and only if the product, $\hat{k}m_{xy}$, is positive, that is, if m_{xy} and $\hat{k}$ are of the same sign. This fact will be shown to be of significance later in the derivation.

5.5.1 Resistance to twisting moment

While no capacity to resist twisting moment is explicitly provided, capacity can be shown to exist on face AB (Figure 5.29a) by considering Equation 5.83, which gives

$$m_{nt}^* = \left(m_y^* - m_x^*\right)\cos\theta\sin\theta$$

from which the excess of capacity over applied moment is

$$g(\theta) = \left[\left(m_y^* - m_y\right) - \left(m_x^* - m_x\right)\right]\cos\theta\sin\theta - m_{xy}(\cos^2\theta - \sin^2\theta) \tag{5.87}$$

However, for $\theta = 0$ and positive m_{xy}, this gives a negative excess capacity; the moment capacity is less than the applied twisting moment. Most bridge decks will have some capacity to resist torsion other than that provided by m_x^* and m_y^*. For example, in reinforced concrete decks, the concrete has capacity to resist shear and torsion other than that provided by reinforcing links/stirrups. In many cases, twisting moments are small, and additional reinforcement to resist the resulting torsion may not be necessary.

5.5.2 New bridge design

When new bridges are being designed, the moment capacities are not generally known in advance, and the problem is one of *selecting* sufficiently large values for m_x^* and m_y^*. It can be seen from Equation 5.86 that m_x^* and m_y^* effectively dictate the critical value, $\hat{k}$, for a particular set of moments; choosing m_x^* and m_y^* amounts to choosing $\hat{k}$. Thus, the designer's problem can be viewed as one of choosing a suitable value for $\hat{k}$ provided that Equation 5.86 is satisfied. It is, of course, also necessary to have positive excess moment capacities.

When choosing a value for $\hat{k}$, it should be borne in mind that there are two roots to Equation 5.86. It can be shown that these roots are related to each other according to

$$\hat{k}_1 = -\frac{1}{\hat{k}_2}$$

where $\hat{k}_1$ and $\hat{k}_2$ are roots 1 and 2, respectively. The simplest choice (not necessarily the least cost choice) is to select $\hat{k}_1 = -\hat{k}_2 = 1$. Equation 5.86 then gives

$$\pm\left[\left(m_y^* - m_y\right) - \left(m_x^* - m_x\right)\right] = 0$$

$$\Rightarrow \left(m_y^* - m_y\right) = \left(m_x^* - m_x\right)$$

that is, the excesses of moment capacity provided in the two orthogonal directions are kept equal. Equation 5.85 gives the excess capacity for m_n:

$$f(\theta) = \left(m_x^* - m_x\right)\cos^2\theta + \left(m_y^* - m_y\right)\sin^2\theta - 2m_{xy}\cos\theta\sin\theta \geq 0$$

As $\cos^2\theta$ is always positive, this can be written as

$$\left(m_x^* - m_x\right) + \left(m_y^* - m_y\right)\hat{k}^2 - 2m_{xy}\hat{k} \geq 0$$

Taking $\hat{k} = \pm 1$, this becomes

$$\left(m_x^* - m_x\right) + \left(m_y^* - m_y\right) \geq 2m_{xy}\hat{k}$$

$$\Rightarrow \left(m_x^* - m_x\right) = \left(m_y^* - m_y\right) \geq m_{xy}\hat{k} \tag{5.88}$$

EXAMPLE 5.2   WOOD AND ARMER EQUATIONS FOR DESIGN

At a point in a bridge slab, the moments per unit length due to applied loads have been found to be $m_x = 80$, $m_y = 190$ and $m_{xy} = 20$. It is required to determine moment capacities to satisfy m_n and m_t.

Selecting, $\hat{k} = 1$ gives Equation 5.89:

$$\left(m_x^* - m_x\right) = \left(m_y^* - m_y\right) \geq \left|m_{xy}\right|$$

$$\Rightarrow \left(m_x^* - 80\right) = \left(m_y^* - 190\right) \geq 20$$

Hence, capacities of $m_x^* = 100$ and $m_y^* = 210$ will be sufficient.

For m_n, the critical value is when the product of m_{xy} and $\hat{k}$ is positive. Hence, if m_{xy} is positive, the critical value for $\hat{k}$ is +1 and if m_{xy} is negative, the critical value for $\hat{k}$ is −1. Hence, Equation 5.88 can be written as

$$\left(m_x^* - m_x\right) = \left(m_y^* - m_y\right) \geq \left|m_{xy}\right| \tag{5.89}$$

Chapter 6

Application of planar grillage and finite element methods

6.1 INTRODUCTION

In Chapter 5, the behaviour of bridge slabs is considered. Two methods of analysis are introduced: grillage and finite element (FE) methods, both of which consist of members lying in one plane only. In this chapter, both of these planar methods of analysis are used to model a range of bridge forms. Planar methods are amongst the most popular currently available for the analysis of slab bridges. They can, with adaptation, be applied to many different types of slab as will be demonstrated. Further, their basis is well understood, and the results are considered to be of acceptable accuracy for most bridges.

In Chapter 7, more complex non-planar methods of analysis are considered. For certain bridges, non-planar models are considerably more accurate than planar models. However, they can also be more complex, and the results can be more difficult to interpret. For this reason, planar grillage and FE models are at present the method of choice of a great many bridge designers for most bridge slabs.

6.2 SIMPLE ISOTROPIC SLABS

When bridge slabs are truly planar, it is a simple matter to prepare a computer model following the guidelines specified in Chapter 5. This will be demonstrated in the following examples.

EXAMPLE 6.1   GRILLAGE MODEL OF TWO-SPAN RIGHT SLAB

A two-span bridge deck is illustrated in Figure 6.1. It is to be constructed of prestressed concrete and is to have a uniform rectangular cross section of 0.8 m depth. The deck is supported on four bearings at either end and on two bearings at the centre, as illustrated in Figure 6.1. A combination of fixed, free-sliding and guided-sliding bearings is used so that the bridge can expand or contract freely in all directions in plane. It is required to design a grillage mesh to accurately represent the deck, given that the concrete has a modulus of elasticity of 35×10^6 kN/m².

Figure 6.2a shows a convenient grillage mesh for this bridge deck. The longitudinal members have been placed along the lines of the bearings, with an additional line at the centre of the deck. As recommended in Section 5.3, a row of longitudinal members has been placed at a distance of 0.3 times the depth from the edge of the slab. The transverse members have been placed at a

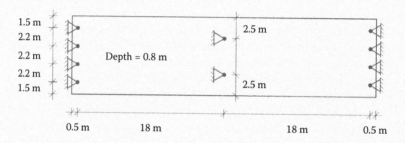

Figure 6.1 Plan view of two-span bridge.

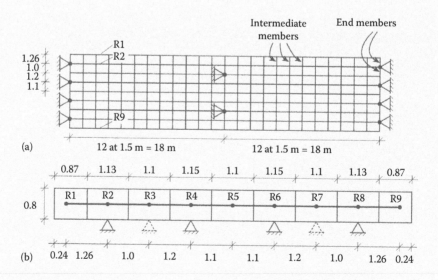

Figure 6.2 Grillage mesh for bridge of Figure 6.1: (a) plan; (b) section.

spacing of 1.5 m, which gives a ratio of transverse to longitudinal spacing of between 1.2 and 1.5. The end rows of transverse members are taken through the centres of the bearings.

Figure 6.2b shows a cross section of the slab with the grillage members superimposed. This is used to determine the breadth of slab attributable to each longitudinal grillage member. It can be seen that this breadth is taken to be from midway between adjacent members on either side. The bridge slab is assumed to be isotropic, and the second moments of area per unit breadth are taken to be equal to those of the slab:

$$i_x^{gril} = i_y^{gril} = \frac{0.8^3}{12} = 0.0427 \ m^4/m$$

The torsion constants per unit breadth are calculated according to Equation 5.56:

$$j_x^{gril} = j_y^{gril} = 2\left(\frac{0.8^3}{12}\right) = 0.0853 \ m^4/m$$

The second moments of area and torsion constants of the grillage members are then determined by multiplying these values by the relevant breadth of each member as taken from Figure 6.2b. These values are presented for all of the grillage members in Table 6.1. The longitudinal members have been grouped by row as R1 to R9, and the transverse members have been grouped as *end members* and *all intermediate members*, as illustrated in Figure 6.2.

For the transverse end members, the breadth is 1.5/2 + 0.5 as the slab extends 0.5 m past the centre of the bearing. However, in keeping with recommendation number 6 of Section 5.3.7, this is reduced by 0.3d = 0.24 m for the calculation of the torsion constant. Similarly, when determining the value of the torsion constant of the longitudinal members in rows R1 and R9, a reduced breadth of (0.87 − 0.24) = 0.63 m was used.

Table 6.1 Grillage member properties for Example 6.1

	Second moment of area (m^4)	Torsion constant (m^4)
Longitudinal members		
R1, R9	0.0371	0.0537
R2, R8	0.0483	0.0964
R3, R7	0.0470	0.0938
R4, R6	0.0491	0.0981
R5	0.0470	0.0938
Transverse members		
End members	0.0534	0.0862
All intermediate members	0.0640	0.1280

EXAMPLE 6.2   FINITE ELEMENT MODEL OF TWO-SPAN RIGHT SLAB

A planar FE model is required for the bridge deck of Example 6.1 and Figure 6.1. Figure 6.3 shows a convenient FE mesh. The breadths of the elements are chosen such that nodes coincide with the locations of the supports. The two rows of elements at each edge of the model could be replaced with one row of 1.5 m breadth, but the extra number of elements in the model chosen is not considered to be excessive. The length of the elements along the span of the bridge was chosen as 1.2 m, which is equal to the breadth of the widest element. This is a somewhat arbitrary choice, and had the length been taken as, say, equal to the average breadth of the elements, a similar degree of accuracy could be expected. As this is an isotropic bridge slab, the only geometric property that has to be assigned to the elements is their depths. All of the elements are assigned a depth of 0.8 m, which is equal to the actual depth of the bridge slab. As for Example 6.1, the elastic modulus is taken to be that of the slab, $E = 35 \times 10^6$ kN/m².

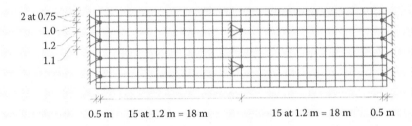

2 at 0.75
1.0
1.2
1.1

0.5 m   15 at 1.2 m = 18 m   15 at 1.2 m = 18 m   0.5 m

Figure 6.3 FE mesh for bridge of Figure 6.1.

6.3 EDGE CANTILEVERS AND EDGE STIFFENING

Slab bridge decks often include a portion of reduced depth at their edges known as an edge cantilever. This type of construction is chosen partly for its reduced self-weight and partly for its slender appearance (see Section 1.8). Cross sections of typical slab decks with edge cantilevers are illustrated in Figure 6.4. Upstands or downstands, such as those illustrated in Figure 6.4c and d, are often included at the edges of the slab, either to stiffen the edge, to carry a protective railing, or simply for aesthetic reasons. These are often important

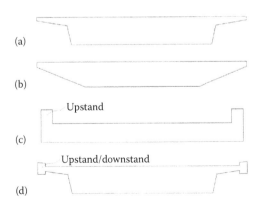

Figure 6.4 Typical cross sections of slab decks showing cantilevers and upstands/downstands.

Figure 6.5 Cross section of slab deck with slender cantilever and upstand.

aesthetically and, in the case of concrete bridges, may be precast to ensure a good quality of finish. In such cases, the upstand may not be integral with the bridge deck and can simply be considered as an additional load on it. If they are made integral with the deck, then the increased stiffness that they provide generally needs to be considered. It is not necessarily conservative to ignore the additional stiffness provided by them.

The effect of an edge cantilever or an integral upstand/downstand is to change the stiffness of the bridge deck. In slab bridges, the appropriate stiffness is determined by first finding the neutral axis location for the complete deck. The properties of each part are then calculated about this axis. In some bridge decks, finding the location of the neutral axis may not be straightforward. Figure 6.5 shows the cross section of a deck with a long, slender edge cantilever, with an upstand at its edge. In such a case, the neutral axis will not remain straight as the upstand tries to bend about its own axis, causing the bridge neutral axis to rise. Bridge decks of this type are discussed further in Chapter 7. Only decks where the neutral axis remains substantially straight are considered here. These will be similar to those illustrated in Figure 6.4, where the edge cantilever is relatively short or stocky, or where the upstand is not excessively stiff. The neutral axis is then taken to be straight across the complete deck and to pass through its centroid.

EXAMPLE 6.3   GRILLAGE ANALYSIS OF SLAB WITH EDGE CANTILEVER

The cross section of a prestressed concrete bridge slab with edge cantilevers is illustrated in Figure 6.6. The bridge deck, which has a constant cross section through its length, spans 20 m and is simply supported on three bearings at each end as indicated in Figure 6.6. It is required to design a suitable mesh of grillage members to model the structure.

The first task is to determine the location of the deck neutral axis, which is taken to be straight, and to pass through the centroid. This can be determined by hand or by using one of many computer programs available for such purposes. In this case, the neutral axis is found to be 563 mm below the top of the bridge deck. Details of a general approach to this calculation are given in Appendix B.

The cross section is divided into a number of segments, each of which is represented by a row of grillage members. Figure 6.7a shows the divisions chosen and the corresponding grillage members. The spacings of longitudinal grillage members are given in Figure 6.7b. The reasons for this particular arrangement are as follows:

- Each edge cantilever is modelled with two separate rows of members so that the reduced depth towards the edge can be allowed for.
- The outermost row of grillage members, Row R1, is placed at a distance of 90 mm from the edge of the cantilever. This distance corresponds to 0.3 times the average depth of cantilever. This is in keeping with recommendation number 6 of Section 5.3.7.
- The second row of grillage members from the edge, R2, is located at the centre of the portion of cantilever which it represents.

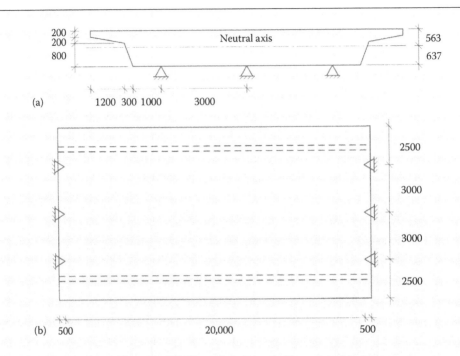

Figure 6.6 Bridge deck of Example 6.3 (dimensions in mm): (a) section; (b) plan.

- The third row of members from the edge, R3, is placed at a distance of 0.3 times the depth of the deck (0.3 × 1200 = 360 mm) from the midpoint of the sloping edge of the main deck. The location from which this distance is taken is subjective, but that chosen here seems reasonable.
- The fourth row, R4, and middle row, R7, of grillage members are located to coincide with the supports to the bridge deck. Note that row R4 is not exactly at the centre of the portion it represents.
- Two rows of grillage members, R5 and R6 (and R8 and R9), are chosen between the supports. In each case, these members represent a portion of bridge slab of breadth 1000 mm, and they are each located at the centre of that portion.

Figure 6.7c illustrates a plan of the grillage mesh with dimensions in millimetres. Twenty one rows of transverse members with a spacing of 1000 mm were chosen. This is a very dense mesh, having a spacing less than the slab depth. However, it gives a good longitudinal-to-transverse-spacing ratio, between 1:1 and 1:1.27. Due to the variation in depth between rows R2 and R3, the transverse members between these rows have been modelled as two separate members with a row of nodes where they join.

For each row of longitudinal grillage members, the second moment of area is calculated about the centroid of the bridge. The second moment of area relative to the centroid of the bridge is always greater than (or equal to) that relative to the centroid of the individual portion of deck. For example, the second moment of area of row R7 is given by

$$I_{R7} = \frac{1(1.2)^3}{12} + (1 \times 1.2)(0.6 - 0.563)^2 = 0.146 \text{ m}^4$$

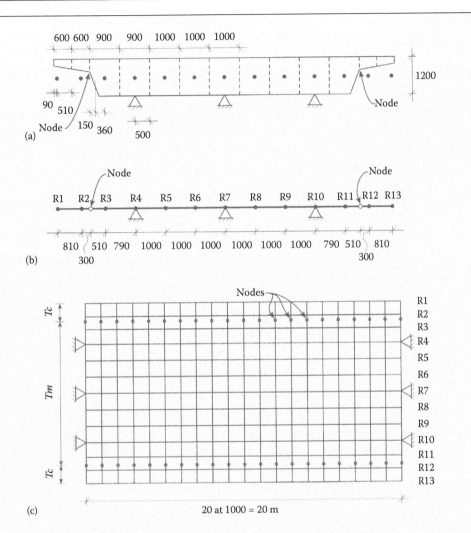

Figure 6.7 Grillage model (dimensions in mm): (a) cross section showing grillage members and corresponding segments of deck; (b) schematic of cross section showing spacing between members; (c) plan of mesh.

All of the longitudinal grillage member second moments of area are presented in Table 6.2.

The transverse members are divided into two groups. The first group are those in the cantilever portion, running from the edge as far as the row of nodes indicated in Figure 6.7. These are labelled Tc in Figure 6.7c. The second group are those in the main portion of the deck and account for all of the other transverse members. These are labelled Tm in the figure.

The second moments of area of the transverse members in the cantilever, Tc, are taken about their own centroids as they will bend (transversely) about their own centroids. The depth of these members is taken as the average depth of the cantilever, that is, 300 mm. The second moment of area per unit breadth of these members is therefore

$$i_{Tc} = \frac{0.3^3}{12} = 0.0023 \text{ m}^3$$

Table 6.2 Grillage member properties for Example 6.3

	Second moment of area (m⁴)	Torsion constant (m⁴)
Longitudinal members		
R1, R13	0.029	0.010
R2, R12	0.034	0.013
R3, R11	0.110	0.143
R4, R10	0.131	0.261
R5, R6, R7, R8, R9	0.146	0.290
Transverse members		
Tc – End members	0.002	0.019
Tc – Intermediate members	0.002	0.021
Tm – End members	0.144	0.178
Tm – Intermediate members	0.144	0.278

The second moments of area of the transverse grillage members in the main part of the deck, Tm, are also calculated about their own centroids as it is about these that they will bend. The second moment of area per unit breadth of these members is therefore

$$i_{Tm} = \frac{1.2^3}{12} = 0.1440 \text{ m}^3$$

The second moment of area of the transverse members is then found by multiplying these values by the breadth of the members (which for all intermediate members is 1 m). The results are presented in Table 6.2.

The torsion constants for the members are determined in accordance with Equation 5.60 as this is an orthotropic deck $\left(i_x^{slab} \neq i_y^{slab} \right)$:

$$j^{gril} = 2\sqrt{i_x^{slab} i_y^{slab}}$$

where i_x^{slab} and i_y^{slab} are the second moments of area per unit breadth in the X and Y directions, respectively. To apply this equation, the X direction is arbitrarily chosen as the longitudinal direction. Considering the longitudinal members in row R1 and the transverse members Tc, the second moment of area per unit breadth of the longitudinal members (with reference to Table 6.2) is given by

$$i_{R1} = \frac{0.029}{0.6} = 0.048 \text{ m}^3$$

The second moment of area per unit breadth of the transverse members is 0.002 m³. Hence, the torsion constant per unit breadth of the longitudinal members, R1, and the transverse members, Tc, is given by

$$j_{R1} = j_{Tc} = 2\sqrt{(0.048)(0.002)} = 0.020 \text{ m}^3$$

Considering next the longitudinal members in row R2 and the transverse members Tc, the second moment of area per unit breadth of the longitudinal members (with reference to Table 6.2) is given by

$$i_{R2} = \frac{0.034}{0.6} = 0.057 \text{ m}^3$$

Therefore, the torsion constant per unit breadth of the longitudinal members, R2, and the transverse members, Tc, is given by

$$j_{R2} = j_{Tc} = 2\sqrt{(0.057)(0.002)} = 0.021 \text{ m}^3$$

This gives a value for the torsion constant per unit breadth for each of the longitudinal members R1 and R2, but there are two distinct values for the transverse members Tc. At this stage, an approximation is made by taking an average value for the torsion constant per unit breadth of the transverse members. In doing this, the condition of Section 5.3.3 is not satisfied, which required that the torques per unit breadth in the grillage members in the longitudinal and transverse directions be of the same magnitude. However, as the two distinct values are very close, the average value is considered acceptable. The torsion constant per unit breadth of the transverse grillage members, Tc, is therefore

$$j_{Tc} = \frac{(0.020 + 0.021)}{2} = 0.021 \text{ m}^3$$

Considering the longitudinal members in row R3 and the transverse members, Tm, the second moment of area per unit breadth of the longitudinal members (with reference to Table 6.2) is given by

$$i_{R3} = \frac{0.110}{0.9} = 0.122 \text{ m}^3$$

The second moment of area per unit breadth of the transverse members is 0.144 m³, and therefore, the torsion constant per unit breadth of the longitudinal members, R3, and the transverse members, Tm, is given by

$$j_{R3} = j_{Tm} = 2\sqrt{(0.122)(0.144)} = 0.265 \text{ m}^3$$

This value is adopted for the longitudinal members in row R3. The other longitudinal members, R4–R10, have the same second moment of area per unit breadth (with reference to Table 6.2), which is

$$i_{R4-R10} = \frac{0.131}{0.9} = \frac{0.146}{1.0} = 0.146 \text{ m}^3$$

Hence, the torsion constant per unit breadth of the longitudinal members, R4–R10, and the transverse members, Tm, is given by

$$j_{R4-R10} = j_{Tm} = 2\sqrt{(0.146)(0.144)} = 0.290 \text{ m}^3$$

This value is adopted for longitudinal members R4–R10. The average of the two values is taken for the transverse members Tm:

$$j_{Tm} = \frac{(0.265 + 0.290)}{2} = 0.278 \text{ m}^3$$

The torsion constant for each grillage member is then arrived at by multiplying the torsion constant per unit breadth by the breadth of slab represented by that member. For the end transverse members, Tm, the breadth is reduced by 0.3 × 1.2 = 0.36 m. For the longitudinal members in rows R3 and R11, the breadth is reduced by 0.3 × 0.9 = 0.27 m. For the end transverse members, Tc, and the longitudinal members in rows R1 and R13, the breadth is reduced by 0.3 × 0.3 = 0.09 m. These values are given in Table 6.2. It can be seen that by splitting the transverse members running between rows R2 and R3 (and R11 and R12) into two separate transverse members, the need to average two dissimilar values of torsion constant was avoided.

EXAMPLE 6.4   FE ANALYSIS OF SLAB WITH EDGE CANTILEVER

It is required to prepare an FE model for the bridge deck of Example 6.3 and Figure 6.6. The cross section of Figure 6.6a is divided into a number of segments in a similar manner to the grillage model. As the nodes form the boundaries of the elements and the location of the supports must coincide with nodes, the division of the deck for the FE model varies somewhat from that of the grillage. Figure 6.8a shows the division of the deck, and Figure 6.8b shows a cross section through the FE model. The depths of the elements have not been drawn to scale in this figure. Figure 6.8c shows a plan of the model with rows of *elements* labelled r1–r14. The length of the elements (in the longitudinal direction) is taken as 1000 mm. This results in 20 elements in each of the 14 longitudinal rows.

The X axis is again chosen to be in the longitudinal direction and the Y axis to be perpendicular to this. The second moments of area per unit breadth, i_x^{slab} and i_y^{slab}, are determined for each portion of the bridge deck. In the X direction, these are calculated about the centroid of the bridge, which was seen in Example 6.3 to be located 563 mm below the top surface. In the Y direction, the second moment of area per unit breadth of each portion is determined about its own centroid as it is about this that transverse bending occurs. In the case of the elements representing the edge cantilevers (rows r1, r2, r13 and r14), the transverse stiffness is based on the average depth of that portion of cantilever. In the case of the elements in row r3, it is difficult to determine the transverse stiffness as the depth varies significantly. A depth of 1000 mm is chosen as this seems to be a reasonable compromise, and it is felt that the problem does not warrant an in-depth analysis. The second moments of area per unit breadth for each row of

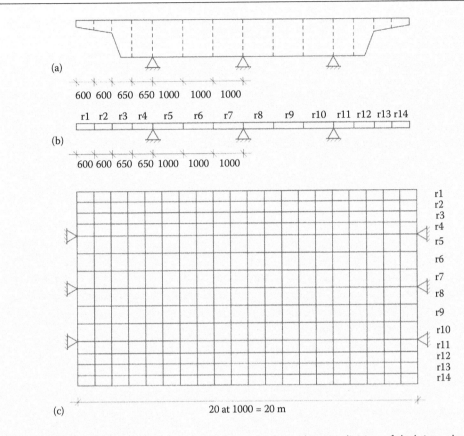

Figure 6.8 FE model (dimensions in mm): (a) cross section showing division of deck into elements; (b) schematic of cross section showing breadths of elements; (c) plan of element mesh.

elements are given in Table 6.3. It is noteworthy that the exact i_x^{slab} values in the cantilever are almost identical to those calculated assuming average thicknesses – 0.0493 and 0.0563 for r1 and r2 respectively.

The bridge deck is geometrically orthotropic as the second moments of area vary in two orthogonal directions. In the FE program, it is modelled as materially orthotropic with a single value for element depth. The variation of the second moment of area in the two directions is allowed for by specifying two different elastic moduli. Arbitrarily choosing the elastic modulus in the X direction, E_x^{elem}, to be equal to the elastic modulus of the concrete, E_c, then the equivalent

Table 6.3 FE properties for Example 6.4

FE row number	i_x^{slab} (m^3)	i_y^{slab} (m^3)	E_x^{elem}	d^{elem} (m)	E_y^{elem}	G_{xy}^{elem}
r1, r14	0.0490	0.0013	E_c	0.838	$0.027\,E_c$	$0.068\,E_c$
r2, r13	0.0561	0.0036	E_c	0.876	$0.064\,E_c$	$0.106\,E_c$
r3, r12	0.1138	0.0833	E_c	1.109	$0.732\,E_c$	$0.356\,E_c$
r4 through r11	0.1456	0.1440	E_c	1.204	$0.989\,E_c$	$0.414\,E_c$

depth, d^{elem}, to be used for the FE is found by equating the second moments of area of the element and the slab (Equation 5.77):

$$i^{elem} = i_x^{slab}$$

$$\Rightarrow \quad d^{elem} = \sqrt[3]{12 i_x^{slab}}$$

Equation 5.78 then gives an expression for the elastic modulus in the Y direction, E_y^{elem}, in terms of the elastic modulus of the concrete, E_c:

$$E_y^{elem} = E_c \frac{i_y^{slab}}{i_x^{slab}}$$

The elastic moduli in the two directions and the equivalent depths of each row of elements are given in Table 6.3.

The shear modulus, G_{xy}^{elem}, is calculated using Equation 5.81 by substituting values for the Poisson's ratio, the elastic modulus and the second moments of area per unit breadth. Assuming a Poisson's ratio of 0.2 for concrete, values of G_{xy}^{elem} were arrived at for each row of elements. These values are also given in Table 6.3.

6.4 VOIDED SLAB BRIDGE DECKS

Longitudinal voids are sometimes incorporated into concrete slab bridge decks to reduce their self-weight while maintaining a relatively large second moment of area. These are created by placing void formers, usually made from polystyrene, within the formwork before casting the concrete. Figure 6.9 shows a cross section through a typical voided slab bridge deck with tapered edges. It is usual to discontinue the voids over the supports, which has the effect of creating solid diaphragm beams there.

When the void diameter is less than approximately 60% of the slab depth, it is common practice to model the voided slab using the same methods as used for solid slab decks. On the other hand, when the void diameter exceeds approximately 60%, the behaviour becomes more 'cellular'. Cellular decks are characterised by the distortional behaviour illustrated in Figure 6.10, which can be modelled using a variation of the conventional grillage or FE methods known as 'shear flexible' grillage or FE. Even if the voids are large, a voided slab deck is less likely to distort than the box girder section of Figure 6.10, and without specific guidance, such a shear flexible model would be difficult to implement. Bakht et al. (1981) reviewed many methods of analysing voided slab bridges. They propose that, regardless of the size of the voids, such slabs can be analysed using the same techniques as those used for solid slab decks but with modified member properties.

The first step in the modelling of a voided slab deck is to determine the location of the centroid. This is generally taken to be at a constant depth transversely and to pass through

Figure 6.9 Cross section through voided slab bridge.

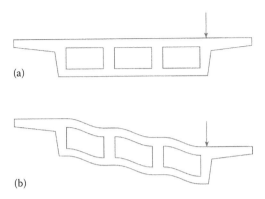

Figure 6.10 Characteristic behaviour of cellular bridge deck: (a) original geometry; (b) deformed shape showing characteristic cell distortion.

the centroid of the deck. If the bridge deck has edge cantilevers or if the voids are not located at the centre, then the position of the centroid may not be at mid-depth and should be calculated in the usual way. For planar grillage or FE models, the properties of each part of the deck are then calculated relative to the centroid of the complete deck.

Determination of the longitudinal second moment of area per unit breadth of a voided slab, i_x^{slab}, is straightforward. The stiffness of the voided portion is simply subtracted from the stiffness of the solid slab. Determination of the transverse second moment of area and the torsional stiffness is not so simple. For the transverse second moment of area, Bakht et al. (1981) recommend using the method of Elliott, which gives this quantity in terms of the depth of the slab, d, and the diameter of the voids, d_v (Figure 6.11):

$$i_y^{v\text{-slab}} = \frac{d^3}{12}\left[1 - 0.95\left(\frac{d_v}{d}\right)^4\right] \tag{6.1}$$

Equation 6.1 does not take into account the spacing of the voids as the authors maintained that this was not a significant factor. Clearly, this equation is only applicable to slabs with a sensible void spacing. A slab where the voids were spaced three to four times the slab depth apart would have a transverse stiffness in excess of that predicted by Equation 6.1. This equation assumes that the centres of the voids and the deck centroid (for transverse bending) are located at mid-depth. This is quite often a reasonable assumption when considering transverse bending.

When the void-diameter-to-slab-depth ratio is 0.6 or less, the transverse stiffness can be approximated as being equal to the longitudinal stiffness. Examination of Equation 6.1 shows that the presence of the voids reduces the transverse stiffness by only 12% for a ratio of 0.6.

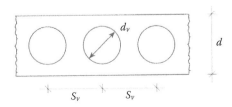

Figure 6.11 Cross section through segment of voided slab bridge.

Table 6.4 Ratio of torsional stiffness of voided slab, $j^{\text{v-slab}}$, to that of solid slab, j^{slab}

		d_v/s_v				
		0.9	0.8	0.7	0.6	0.5
$\dfrac{d_v}{d}$	0.90	0.45	0.48	0.51	0.56	0.62
	0.85	0.55	0.58	0.61	0.64	0.69
	0.80	0.64	0.66	0.68	0.71	0.75
	0.75	0.70	0.72	0.74	0.77	0.80
	0.70	0.76	0.78	0.79	0.82	0.84
	0.65	0.81	0.82	0.84	0.86	0.88
	0.60	0.85	0.86	0.87	0.89	0.90

Source: Bakht, B. et al., *Canadian Journal of Civil Engineering*, 8, 376–391, 1981.

For the torsional stiffness of voided slabs per unit depth, $j^{v\text{-slab}}$, Bakht et al. (1981) recommend using the method of Ward and Cassell. This gives the values presented here in Table 6.4 for the ratio of torsional stiffness of the voided slab $j^{v\text{-slab}}$ to that of a solid slab of the same depth, j^{slab}. For a grillage model, j^{slab} can be determined from Equation 5.55 or 5.60, and Table 6.4 can then be used to determine $j^{v\text{-slab}}$. It was suggested that the values given in Table 6.4 are only applicable to internal voids in an infinitely wide slab because those at the edges possess much lower torsional stiffness. However, Bakht et al. (1981) conclude that, in most practical cases, reduction of the torsional stiffness for the edge voids is not warranted as voided slab bridge decks are usually tapered at their edges or have substantial edge cantilevers.

EXAMPLE 6.5   GRILLAGE MODEL OF VOIDED SLAB BRIDGE

Figure 6.12 shows the cross section of a prestressed concrete bridge deck, which incorporates circular voids along its length. The deck spans 24 m between the centres of supports and is supported on four bearings at either end, as illustrated in Figure 6.12. The voids stop short at each end, forming solid diaphragm beams 1 m wide over the supports. Thus, the total bridge is 25 m long, consisting of 23 m of voided section and two 1 m diaphragms. The centroid of the deck is located at mid-depth as the voids are located there. The layout and member properties are required for a grillage model.

Figure 6.13 shows a suitable grillage mesh. The longitudinal members are located midway between voids, with the exception of the outer row on each side where they are located

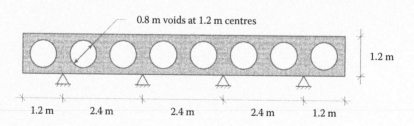

Figure 6.12 Cross section through bridge of Examples 6.5 and 6.6.

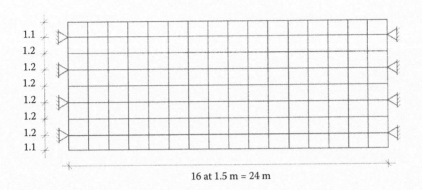

1.1
1.2
1.2
1.2
1.2
1.2
1.2
1.1

16 at 1.5 m = 24 m

Figure 6.13 Grillage mesh for bridge of Example 6.5.

midway between the edge of the outermost void and the edge of the deck. It is not considered appropriate to locate these grillage members at 0.3 times the depth of the slab from the edge as this location is within the void. By using this arrangement, the supports coincide with the locations of nodes in the grillage mesh. The transverse grillage members are located in 17 rows, 1.5 m apart.

As the void diameters are in excess of 60% of the slab depth, the slab is treated as an orthotropic plate, and the properties of the longitudinal and transverse members are determined separately. The longitudinal direction is taken to be the X direction. The internal longitudinal grillage members represent the portion of deck illustrated in Figure 6.14. The second moment of area of this member is found by subtracting the second moment of area of the circle from that of the rectangle:

$$I_x = \frac{1.2(1.2)^3}{12} - \frac{\pi(0.8)^4}{64} = 0.153 \ \text{m}^4$$

The edge longitudinal grillage member represents a portion of deck equal to exactly half that of the internal members, with the result that its second moment of area is given by

$$I_x = \frac{0.153}{2} = 0.076 \ \text{m}^4$$

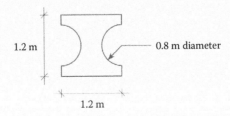

1.2 m

0.8 m diameter

1.2 m

Figure 6.14 Segment of voided slab.

The second moments of area of the internal transverse members are determined using Equation 6.1:

$$i_y^{\text{v-slab}} = \frac{d^3}{12}\left[1 - 0.95\left(\frac{d_v}{d}\right)^4\right]$$

$$= \frac{1.2^3}{12}\left[1 - 0.95\left(\frac{0.8}{1.2}\right)^4\right]$$

$$= 0.117\,\text{m}^3$$

Hence, for the internal transverse members, the second moment of area is

$$I_y = 1.5(0.117) = 0.176\ \text{m}^4$$

For the 1 m wide end diaphragms, the second moment of area is simply

$$I^{\text{diaph}} = 1(1.2)^3/12 = 0.144\ \text{m}^4$$

As the diaphragm is only 1 m wide and the transverse members are spaced at 1.5 m, the next row of transverse members, adjacent to the diaphragm, will be 1.75 m wide and will have a second moment of area of

$$I_y = 1.75(0.117) = 0.205\ \text{m}^4$$

The torsion constant for the grillage members is found from Table 6.4. Both the ratio d_v/s_v and d_v/d are 0.67. Interpolating in Table 6.4 gives a ratio for the torsion constants per unit breadth of

$$\frac{j^{\text{v-slab}}}{j^{\text{slab}}} = 0.83$$

Taking Equation 5.55 to calculate the torsion constant per unit breadth for a solid slab then gives

$$j^{\text{v-slab}} = 0.83 \times d^3/6$$

$$= 0.83 \times (1.2^3/6)$$

$$= 0.239\,\text{m}^4/\text{m}$$

The torsion constants for both the longitudinal and transverse members in the voided slab are then found by multiplying this value by their respective breadths. The torsion constant per unit breadth for the diaphragm is given by Equation 5.60:

$$j^{\text{diaph}} = 2\sqrt{i_x^{\text{v-slab}} i^{\text{diaph}}}$$

$$= 2\sqrt{\left(\frac{0.153}{1.2}\right)\left(\frac{0.144}{1.0}\right)}$$

$$= 0.271\,\text{m}^4/\text{m}$$

$$\Rightarrow \quad j^{\text{diaph}} = 0.271\,\text{m}^4$$

EXAMPLE 6.6   FINITE ELEMENT MODEL OF VOIDED SLAB BRIDGE

An FE model is required for the 25 m long voided slab deck of Example 6.5 and Figure 6.12.

For convenience, a mesh consisting largely of 1.2 m square elements is chosen, as illustrated in Figure 6.15. At the ends, two transverse rows of elements, each 0.5 m wide, are used to represent the diaphragm. The transverse rows of elements adjacent to the diaphragms at each end are 1.3 m wide in order to make up the correct total length.

Each longitudinal row of elements represents a strip of the deck from midway between two voids to midway between the next two. The second moment of area per unit breadth in the longitudinal direction can be found by considering a 1.2 m wide strip of the deck. The total second moment of area of this strip is again calculated by subtracting the second moment of area of the void from that of the equivalent rectangular section:

$$I_x^{\text{v-slab}} = \frac{1.2(1.2)^3}{12} - \frac{\pi(0.8)^4}{64} = 0.153 \text{ m}^4$$

Hence, the second moment of area per unit breadth is

$$i_x^{\text{v-slab}} = \frac{0.153}{1.2} = 0.127 \text{ m}^3$$

For the transverse direction, Equation 6.1 gives

$$i_y^{\text{v-slab}} = \frac{1.2^3}{12}\left[1 - 0.95\left(\frac{0.8}{1.2}\right)^4\right] = 0.117 \text{ m}^3$$

The slab is geometrically orthotropic as the second moments of area (rather than the moduli of elasticity) are different for the longitudinal and transverse directions. To model this as a materially orthotropic plate, it is necessary to calculate a single equivalent value for slab depth, d_e. Selecting the modulus of elasticity in the X direction, E_x, equal to the modulus for the concrete, then Equation 5.77 implies a depth of element of

$$d_e = \sqrt[3]{12 i_x^{\text{v-slab}}} = \sqrt[3]{12(0.127)} = 1.15 \text{ m}$$

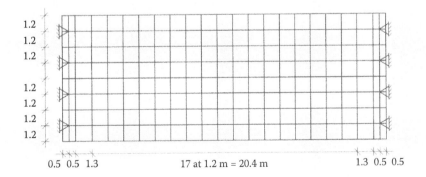

Figure 6.15 FE mesh for bridge of Example 6.6.

Equation 5.78 gives an expression for the elastic modulus in the Y direction:

$$E_y^{elem} = E^{v\text{-slab}}\, \frac{i_y^{v\text{-slab}}}{i_x^{v\text{-slab}}} = E^{v\text{-slab}}\left(\frac{0.117}{0.127}\right) = 0.921 E^{v\text{-slab}}$$

where $E^{v\text{-slab}}$ is the modulus of elasticity of the concrete in the voided slab. The shear modulus is calculated from Equation 5.81:

$$G_{xy}^{elem} = \frac{E^{v\text{-slab}}}{2(1+v)}\sqrt{\frac{i_y^{v\text{-slab}}}{i_x^{v\text{-slab}}}}$$

Taking a Poisson's ratio of 0.2, this gives

$$G_{xy}^{elem} = \frac{E^{v\text{-slab}}}{2(1.2)}\sqrt{\frac{0.117}{0.127}}$$

$$= 0.4 E^{v\text{-slab}}$$

The diaphragm beams are solid; thus the corresponding elements are 1.2 m thick and have moduli of elasticity in both directions equal to that of the concrete. The shear modulus for the diaphragms is given by Equation 5.57.

6.5 BEAM-AND-SLAB BRIDGES

Beam-and-slab decks (also known as girder bridges) are used for a wide range of medium spans. They differ from slab bridge decks in that a large portion of their stiffness is concentrated in discrete beams, which run in the longitudinal direction. Load sharing between the beams may be provided by a top slab or by a combination of a top slab and a number of transverse diaphragm beams. Beam-and-slab bridges are generally suitable for similar span lengths as slab bridges but are often chosen in preference because of their ability to be easily erected over inaccessible areas such as deep valleys or live roads or railways.

Beam-and-slab decks may be formed in a number of ways, the most obvious being the casting of an in situ concrete slab on steel or precast concrete beams, as shown in Figure 6.16a

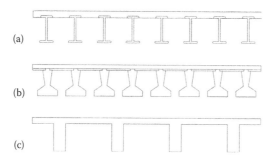

Figure 6.16 Forms of beam-and-slab construction: (a) in situ slab on steel beams; (b) in situ slab on precast concrete beams; (c) in situ beam and slab.

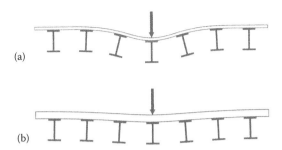

Figure 6.17 Load sharing in beam-and-slab decks: (a) thin slab – little load sharing; (b) thick slab – increased load sharing.

and b. Many other methods exist, such as steel beams with a composite steel and concrete slab, a precast concrete slab or even a completely in situ beam and slab, as illustrated in Figure 6.16c.

During construction, the beams generally act alone and must be capable of carrying their self-weight, the weight of the slab and any construction loads present. On completion, the structural action of these decks is considered to be two-dimensional. Therefore, they can be analysed by similar methods to those proposed for slab decks in the preceding sections. The main load-carrying component of a beam-and-slab deck is the longitudinal spanning beams.

The slab spans transversely between beams and transmits applied loads to them but also acts as a flange to those longitudinal beams. In addition to this, the slab provides a means for load sharing between beams. The extent of this load sharing is largely dependent on the stiffness of the slab. Consequently, it is important that the slab be idealised correctly in the model as, for example, an overly stiff slab may lead to a prediction of load sharing between adjacent beams, which does not occur in reality. This phenomenon is illustrated in Figure 6.17.

Transverse diaphragm beams can be used to provide additional load sharing between longitudinal beams. Wide diaphragms also serve to improve the shear capacity by extending the portion of the bridge near a support which is solid. In precast concrete beam construction, continuity between adjacent spans may be provided by the slab alone, but quite often, a diaphragm beam is constructed over intermediate supports to provide additional continuity.

6.5.1 Grillage modelling

Grillage modelling of beam-and-slab decks generally follows the same procedures as for slab decks. The obvious exception is that grillage beams should normally be positioned at the location of the longitudinal beams. This generally complies with the need to locate beams at the supports as, in beam-and-slab construction, supports are normally provided directly beneath the beams. It is possible to use one grillage member to represent two or more actual beams, but this complicates the calculation of properties and the interpretation of results, with little saving in analysis time in most cases.

The properties of the longitudinal grillage members are determined from the properties of the actual beams and the portion of slab above them. Unlike slab decks, the section properties for beam-and-slab decks are generally calculated about the centroid of this composite section, not about the centroid of the whole bridge. This approach is justified on the basis that, due to the low stiffness of the slab, there will be a much greater variation in the depth of the centroid than in slab bridges.

Transverse grillage members should clearly be placed at the location of all diaphragm beams. The slab will act as a flange to such beams, making them T- or L-section in shape. Hambly (1991) suggests an effective flange breadth of $b_w + 0.3s$ for L-sections, as illustrated in Figure 6.18, where s is the spacing between beams. Transverse members are also required to represent the transverse stiffness of the slab. For slab decks, Section 5.3.7 stated that transverse member spacing should be between one and three times the longitudinal member spacing. This spacing is also recommended for beam-and-slab bridges, although greater spacings are possible without significant loss of accuracy. The properties of the transverse grillage members should be derived from the properties of the relevant diaphragm beam or slab as appropriate, each acting about its own axis.

If the web width at the top of the longitudinal beams in a beam-and-slab deck is large relative to their spacing, then the slab can inadvertently be modelled as having an excessively long transverse span. Figure 6.19a shows a deck consisting of a concrete slab on precast concrete U-beams. Figure 6.19b shows a grillage model with longitudinal grillage beams for the U-beams and transverse beams spanning between them representing the slab. It can be seen from this that the span of the slab in the model is too long. This would lead to an excessively flexible slab, which in turn would lead to the incorrect modelling of load sharing between the U-beams. One possible solution to this is shown in Figure 6.19c, where the transverse grillage members have been subdivided to include much stiffer portions at their ends.

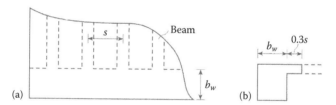

Figure 6.18 Effective flange width of diaphragm beam: (a) plan at end from above; (b) section through L-beam.

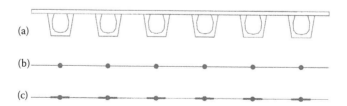

Figure 6.19 Transverse modelling of decks with wide flanges: (a) in situ slab on precast concrete U-beams; (b) conventional grillage model where slab has excessive transverse span; (c) improved grillage model.

EXAMPLE 6.7 GRILLAGE MODEL OF BEAM-AND-SLAB BRIDGE

Figure 6.20 shows the cross section of a beam-and-slab bridge deck consisting of a cast in situ slab on precast concrete Y-beams (Taylor et al. 1990). Each precast beam is supported on a bearing at each end, and the deck has a single span of 20 m (centre to centre of bearings). Solid diaphragm beams, 1 m wide, are provided at each end, and no additional transverse beams are located between these. The elastic modulus of the precast beams is 34 kN/mm² and that of the in situ slab is 31 kN/mm². A grillage model of the beam-and-slab deck is required.

The modular ratio for the in situ and precast concrete is

$$m = \frac{31}{34} = 0.91$$

The procedure adopted is to assign a modulus of elasticity of 34 kN/mm² to all of the grillage members (except for the end diaphragms), but to factor the stiffness of the slab by this modular ratio. The section properties of the precast beam are generally given by the manufacturer. In this case, the properties are

Area = 0.374 m²
Second moment of area = 0.0265 m⁴
Height of centroid above soffit = 0.347 m

The torsion constant may need to be calculated. Figure 6.20b shows the exact dimensions of the precast beam. For the purposes of determining the torsion constant, the beam cross section is approximated as two rectangles, as illustrated in Figure 6.21. The torsion constant of a cross section made up of rectangles is commonly estimated by calculating the torsion constants of the individual rectangles and summing. The torsion constant, J, for a rectangular section according to Ghali et al. (2009) is

$$J = ba^3 \left[\frac{1}{3} - 0.21 \frac{a}{b} \left(1 - \frac{a^4}{12b^4} \right) \right] \tag{6.2}$$

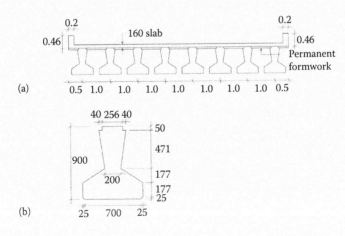

Figure 6.20 Beam-and-slab bridge deck: (a) cross section; (b) detailed dimensions of Y-beam.

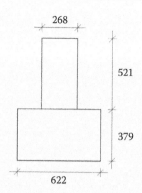

Figure 6.21 Equivalent section made up of rectangles for determination of torsion constant.

where *b* is the length of the longer side and *a* is the length of the shorter side. Applying this equation to the rectangles of Figure 6.21 gives a torsion constant for the Y-beam of

$$J = (0.521)(0.268)^3 \left[\frac{1}{3} - 0.21 \frac{0.268}{0.521} \left(1 - \frac{0.268^4}{12(0.521)^4} \right) \right]$$
$$+ (0.622)(0.379)^3 \left[\frac{1}{3} - 0.21 \frac{0.379}{0.622} \left(1 - \frac{(0.379)^4}{12(0.622)^4} \right) \right]$$

$$\Rightarrow \quad J = 0.0093 \text{ m}^4$$

The constant can be found more exactly by applying Prandtl's membrane analogy as described by Timoshenko and Goodier (1970). A finite difference technique was used to determine the constant in this case and it was found to be $J = 0.0100$ m^4. The simplified method can be seen to be accurate to within 7% for this section.

Figure 6.22 shows a suitable grillage layout for this bridge deck. A longitudinal grillage member is positioned at the location of each Y-beam. Transverse members are positioned at each end to model the diaphragms. Additional transverse beams are located at 2 m centres between these to represent the transverse stiffness of the slab. This gives a transverse-to-longitudinal member spacing

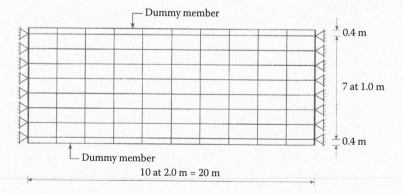

Figure 6.22 Plan view of grillage mesh.

ratio of 2:1, which is acceptable. 'Dummy' longitudinal members with nominal stiffness are provided at the edges, and transverse members are continued past the ends of the edge Y-beams to connect to them. This is a convenient method for applying loads such as those due to parapet railings. Some grillage programs allow the definition of 'dummy' beams. If this is not the case, then these beams should be assigned very small section properties relative to those used elsewhere in the grillage (say, 0.5%). Supports are located at the ends of each longitudinal beam (other than the dummy beams). As the grillage model is planar, consideration need not be given to in-plane horizontal movements at this stage. For the interior longitudinal members, the second moment of area is the sum of the second moment of area of the Y-beam plus the 1 m width of slab above it, both taken about the common centroidal axis of the section. The stiffness of the slab is reduced by factoring it by the modular ratio. Hence, the equivalent area of the combined section is

$$A_{eq} = 0.374 + 0.91(1.0 \times 0.16) = 0.52 \text{ m}^2$$

The section centroid is found by summing moments of area about the soffit:

$$(0.374)(0.347) + 0.91(1.0 \times 0.16)(0.98) = y_b A_{eq}$$

where y_b is the distance of the centroid above the soffit. Hence,

$$y_b = 0.524 \text{ m}$$

The second moment of area of the combined section is

$$I = 0.0265 + 0.374(0.524 - 0.347)^2 + 0.91(1)(0.16)^3/12$$
$$+ 0.91(0.16)(0.98 - 0.524)^2$$

$$\Rightarrow \quad I = 0.0688 \text{ m}^4$$

The torsion constant is taken as the sum of the torsion constants of the Y-beam and the slab. The torsion constant of the slab is determined using Equation 5.55. Hence,

$$J = 0.0100 + 1(0.16)^3/6 = 0.0107 \text{ m}^4$$

Each edge longitudinal member is similar to the interior members except for a 0.2×0.3 m^2 upstand. This raises the centroid above that for the interior members. Summing moments of area about the soffit gives

$$0.520(0.524) + 0.91(0.2 \times 0.3)(1.21) = y_b^{edge}(0.520 + 0.91[0.2 \times 0.3])$$
$$\Rightarrow \quad y_b^{edge} = 0.589 \text{ m}$$

Hence, the second moment of area of the edge section is

$$I_{edge} = 0.0688 + 0.520(0.589 - 0.524)^2 + 0.91(0.2 \times 0.3^3/12)$$
$$+ 0.91(0.2 \times 0.3)(1.21 - 0.589)^2$$

$$\Rightarrow \quad I_{edge} = 0.0925 \text{ m}^4$$

For the transverse members, the properties are determined in the usual manner. For the second moment of area

$$I = \frac{0.91(2)(0.16)^3}{12} = 0.00062 \text{ m}^4$$

The torsion constant is

$$J = \frac{0.91(2)(0.16)^3}{6} = 0.00124 \text{ m}^4$$

The slab acts as a flange to the diaphragm beams. The recommended flange breadth is the sum of the web breadth plus 0.3 times the beam spacing:

$$b_f = 1.0 + 0.3(1.0) = 1.3 \text{ m}$$

Hence, the centroid is

$$\frac{(1 \times 1.06)(0.53) + (0.3 \times 0.16)(0.98)}{(1 \times 1.06) + (0.3 \times 0.16)} = 0.549 \text{ m}$$

above the soffit. For the slab bending about its own axis, the row of transverse members adjacent to the diaphragm accounts for the slab up to 1 m from the centre of the diaphragm, as illustrated in Figure 6.23.

This leaves 0.5 m of slab to be accounted for in the diaphragm stiffness, 0.2 m of which is deemed to be bending about its own axis. The second moment of area is thus

$$\begin{aligned} I_{diaph} &= \frac{1 \times 1.06^3}{12} + (1 \times 1.06)(0.549 - 0.530)^2 \\ &+ \frac{0.5 \times 0.16^3}{12} + (0.3 \times 0.16)(0.98 - 0.549)^2 \\ &= 0.109 \text{ m}^4 \end{aligned}$$

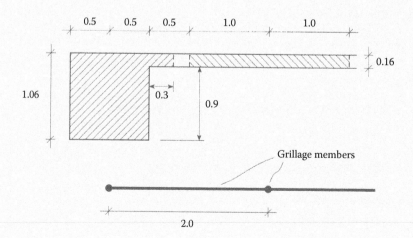

Figure 6.23 Section through end diaphragm beam.

The torsion constant is calculated allowing for 0.5 m of flange from Equations 5.55 and 6.2:

$$J_{diaph} = (1.06)(1.0)^3 \left[\frac{1}{3} - 0.21 \frac{1.0}{1.06} \left(1 - \frac{1.0^4}{12(1.06)^4} \right) \right] + \frac{0.5 \times 0.16^3}{6}$$

$$= 0.158 \ m^4$$

The modulus of elasticity for in situ concrete is used for the diaphragm beams.

6.5.2 Finite element modelling

In FE modelling of beam-and-slab decks, a combined model is generally used, which represents the slab with FEs and the composite beams with beam members. This is generally straightforward to implement and follows the recommendations made for slab bridge decks. Care should be taken when determining the properties of the FEs representing the slab. One of two approaches can be taken. In the first approach, the slab is modelled using isotropic elements, which are assigned a thickness equal to the depth of the actual slab. They are also assigned the elastic properties of the slab. The longitudinal grillage members are then assigned the stiffnesses of the combined beam and associated portion of slab minus those already provided through the FEs. In the second approach, the slab is modelled using ortho-tropic FEs with the true transverse and longitudinal properties applied in both directions. The beams are then modelled by grillage members with the properties of the actual beams, excluding the contribution of the slab.

EXAMPLE 6.8   FINITE ELEMENT MODEL OF BEAM-AND-SLAB BRIDGE

An FE model is required for the beam-and-slab bridge of Example 6.7 and Figure 6.20.

Figure 6.24 shows a suitable FE mesh incorporating beam members longitudinally. Beam members are used for each of the Y-beams and for each of the end diaphragms. The FEs continue to the edge of the deck, resulting in a row of elements 0.5 m wide at each side. An element length of 1 m in the longitudinal direction results in a maximum element aspect ratio of 1:2, which is considered to be acceptable. Supports are provided at the ends of each longitudinal beam member.

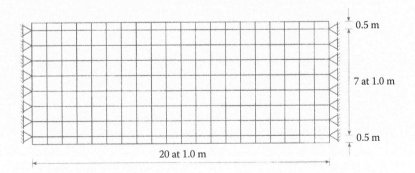

Figure 6.24 Combined FE and beam mesh.

The FEs are assigned a thickness of 0.16 m, which is equal to the depth of the slab. They are assigned a modulus of elasticity and a Poisson's ratio equal to those of the concrete in the slab. For the longitudinal beam members, the properties of the combined Y-beam and the 1 m width of slab above it are determined relative to the centroidal axis of the combined section. The stiffness of the slab, which has already been applied through the FE, is excluded. The modulus of elasticity and Poisson's ratio for the beams are used for these members. Hence, referring to Example 6.7, the second moment of area of the combined section is

$$I_{comb} = 0.0265 + 0.374(0.524 - 0.347)^2 + 0.91(0.16)(0.98 - 0.524)^2$$
$$= 0.0685 \text{ m}^4$$

The torsion constant for the combined section was found in Example 6.7 by adding the individual torsion constants of the Y-beam and slab. As the slab is represented by the elements, the torsion constant for the beam members in this case is simply that of the Y-beam:

$$J = 0.0100 \text{ m}^4$$

In Example 6.7, the second moment of area for the end diaphragms in the grillage model was calculated as (refer to Figure 6.23)

$$I_{diaph} = \frac{1 \times 1.06^3}{12} + (1 \times 1.06)(0.549 - 0.530)^2$$
$$+ \frac{0.5 \times 0.16^3}{12} + (0.3 \times 0.16)(0.98 - 0.549)^2$$

For the FE model, the elements are present up to the centre of the diaphragm to represent the transverse stiffness of the slab about its own axis. Hence, the stiffness of the slab bending about its own axis is not required, and the small component of stiffness inadvertently contributed by the elements should be subtracted:

$$I_{diaph} = \frac{1 \times 1.06^3}{12} + (1 \times 1.06)(0.549 - 0.530)^2$$
$$- \frac{0.5 \times 0.16^3}{12} + (0.3 \times 0.16)(0.98 - 0.549)^2$$
$$= 0.108 \text{ m}^4$$

The torsion constant is that of a rectangular section less the portion inadvertently added through the elements. From Equations 5.55 and 6.2

$$J_{diaph} = (1.06)(1.0)^3 \left[\frac{1}{3} - 0.21 \frac{1.0}{1.06} \left(1 - \frac{1.0^4}{12(1.06)^4} \right) \right] - \frac{0.5 \times 0.16^3}{6}$$
$$= 0.157 \text{ m}^4$$

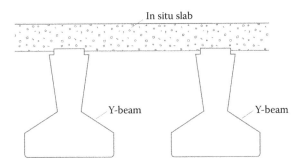

Figure 6.25 Detail of section in beam-and-slab deck.

6.5.3 Transverse local behaviour of beam-and-slab bridges

The top slab in a beam-and-slab bridge is sometimes designed as a one-way slab, spanning transversely between the longitudinal beams (Figure 6.25). However, such an approach results in a great quantity of reinforcement and has been shown to be quite conservative (the AASHTO code mentions a factor of safety for working stresses of about 10 relative to tests). The beams have a considerable lateral stiffness and have the effect of confining the slab. The result is that load is transferred from the slab to the beams largely by 'compressive membrane' or 'internal arching' action rather than bending action alone. In two reported cases (Bakht and Jaeger 1997), Canadian bridges have been built without any transverse slab reinforcement but using steel straps to guarantee confinement. In these cases, the slab-depth-to-beam-spacing ratios were 1:12 and 1:13.5.

To account for observed arching action, the Ontario Highway Bridge Design Code (OHBDC 1992) introduced a provision in the 1990s to allow much less reinforcement than would be found by an assumption of bending. Today, in the United Kingdom, the Design Manual for Roads and Bridges document (BD81/02 2002) describes a procedure for the assessment of bridges using arching action. The American AASHTO code (2010) also describes an 'empirical design' approach in Section 9.7.2 that takes account of this phenomenon.

6.6 CELLULAR BRIDGES

Cellular bridge decks are formed by incorporating large voids within the depth of the slab. The most common type is box girder, with single or multiple rectangular cells. Voided slab bridges, with large diameter circular voids, can also be considered as of a cellular form. However, as was discussed in Section 6.4, alternative methods are available for their analysis, which are generally more convenient. Figure 6.26 shows a number of commonly used cellular deck forms.

There are four principal forms of structural behaviour associated with cellular bridges. The first two of these are longitudinal and transverse bending, as illustrated in Figure 6.27a and b. The third form of behaviour is twisting as indicated in Figure 6.27c. The fourth form, which characterises cellular structures, is transverse cell distortion, as indicated in Figure 6.27d. This distortion is caused by the localised bending of the webs and flanges of the individual cells. The behaviour is similar to that observed in Vierendeel girders.

It is the transverse distortional behaviour that makes the analysis of cellular decks different from other forms. The principal factors affecting the distortion are the dimensions of the cells relative to the deck depth, the stiffness of the individual webs and flanges and the extent

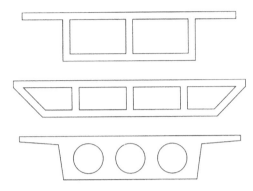

Figure 6.26 Sections through alternative cellular bridge decks.

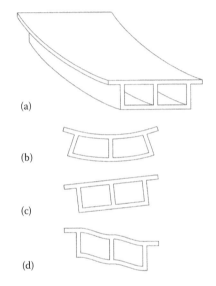

Figure 6.27 Behaviour of cellular decks: (a) longitudinal bending; (b) transverse bending; (c) twisting; (d) transverse distortion.

(if any) of transverse bracing to the cells. Clearly, the provision of transverse diaphragms along the span of a cellular deck will significantly reduce the transverse distortion.

6.6.1 Grillage modelling

Grillage modelling of cellular bridge decks can be achieved with what is commonly referred to as a 'shear flexible' grillage. In this method, the deck is idealised as a grillage of beam members in the usual manner, except that the transverse members are given a reduced shear area designed to allow a shear distortion equal to the actual transverse distortion of the cells in the bridge deck. Clearly, such a method requires a grillage program which models shear deformation as well as bending, and which allows for the specification of a shear (or 'reduced') area for the members independently of the other section properties. The method is illustrated here by means of an example.

Figure 6.28a shows a single cell of width l of a cellular bridge deck under the action (transversely) of a vertical load P. If it were assumed for now that the webs are stiff and that

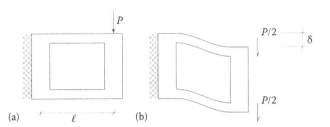

Figure 6.28 Distortion of single cell with stiff webs: (a) applied loading; (b) distorted shape.

transverse distortion is caused by bending of the flanges only, then the distorted shape of the cell is as shown in Figure 6.28b. If the flanges are of equal thickness, then the load acting on each can be taken as $P/2$.

The vertical deflection due to the bending of a beam of length l, fixed against rotation at both ends and subjected to a vertical force $P/2$, is

$$\delta = \frac{(P/2)l^3}{12EI} = \frac{Pl^3}{24EI} \tag{6.3}$$

where I is the second moment of area, and E is the modulus of elasticity. The flanges of a cell will act as beams transversely with a second moment of area per unit breadth equal to $d^3/12$, where d is their thickness. Hence, from Equation 6.3, the deflection due to flange distortion is

$$\delta = \frac{Pl^3}{2Ed^3} \tag{6.4}$$

The total deflection in a cantilever of length l subjected to a vertical load per unit breadth of P at its free end is

$$\delta = \frac{Pl^3}{3EI} + \frac{Pl}{Ga_s} \tag{6.5}$$

where G is the shear modulus, and a_s is the shear area of the section per unit breadth. The second term is the deflection due to shear deformation, which, for most structures, is small relative to the deflection due to bending. By equating the shear deformation in a transverse grillage member to the bending deformation of the cell flanges in the bridge, an expression for the required shear area per unit breadth of a shear flexible grillage member is found:

$$a_s = \frac{2Ed^3}{Gl^2} \tag{6.6}$$

In this example, it was assumed that transverse distortion was caused by the distortion of the cell flanges only. In practice, the webs of cellular decks are also flexible, and consequently, they too contribute to the overall transverse distortion.

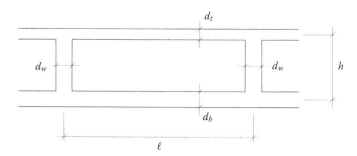

Figure 6.29 Cross section through cellular deck showing dimensions of cell.

Figure 6.29 shows a single cell of a cellular bridge deck with a constant web thickness but different upper and lower flange thicknesses. Assuming points of contraflexure at mid-height and equating the deflection of this cell to the shear deformation of a grillage member give a more exact and general expression for shear area per unit breadth:

$$a_s = \left(\frac{E}{G} \right) \left(\frac{d_w^3 \left(d_t^3 + d_b^3 \right)}{lh \left(d_t^3 + d_b^3 \right) + l^2 d_w^3} \right) \tag{6.7}$$

Details of the derivation of this formula are given in Appendix C. For cellular decks of other shapes, it has been suggested that a plane frame analysis be carried out to determine the equivalent shear area of the transverse grillage members. However, this may be difficult to carry out accurately in practice due to such factors as cracking in concrete sections. (A better solution would be a three-dimensional FE analysis using planar elements with six degrees of freedom per node. Such a model would capture both global and local behaviours simultaneously.)

The second moments of area of the longitudinal members in a shear flexible grillage are determined in the same way as for slab or beam-and-slab decks. For bridges that are transversely stiff or not very wide, a common centroid might be assumed. Then, the centroid of the bridge deck is first determined, and the second moment of area of the portion of deck represented by each longitudinal grillage member is determined about that axis. For wider bridges or bridges that are transversely flexible, the second moment of area for each part of the deck may be calculated about its own centroid.

For the transverse members, the second moment of area of the top and bottom flanges is calculated about the transverse centroidal axis. This can be calculated by summing moments of area about the centroid. Referring to Figure 6.30

$$d_t y_t = d_b y_b$$

which gives the relative values of d_t and d_b. The second moment of area is then

$$i^{\text{trans}} = \frac{d_t^3}{12} + d_t y_t^2 + \frac{d_b^3}{12} + d_b y_b^2 \tag{6.8}$$

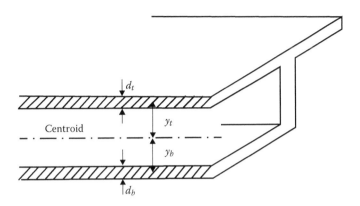

Figure 6.30 Longitudinal section through deck for transverse bending.

The $d_t^3/12$ and $d_b^3/12$ terms in this equation are generally small relative to the others and are often ignored.

The torsion constants of the longitudinal and transverse grillage members are based on the portion of section represented by the members. As mentioned previously, the torsion constant for a thin rectangular section twisting about its own axis may be approximated by $bd^3/3$, where b is the breadth and d is the thickness. Such an equation is valid when the shear flows are opposing through the depth of the section, as illustrated in Figure 6.31a. For a portion of box section, this is not the case, as illustrated in Figure 6.31b, except in the edge cantilevers. The torsion constant for a thin-walled box section is given by

$$J = \frac{4a^2}{\sum_i \left(\dfrac{l_i}{d_i} \right)}$$

(6.9)

where a is the area enclosed by the centre line of the wall, l_i, is an increment of length and d_i is the thickness of that increment. Applying Equation 6.9 to the single cell of Figure 6.29 would give

$$J_{cell} = \frac{4(bl)^2}{\left(\dfrac{l}{d_t} + \dfrac{l}{d_b} + 2\dfrac{b}{d_w} \right)}$$

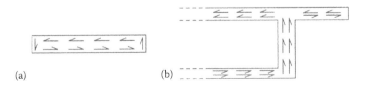

(a)   (b)

Figure 6.31 Shear stresses due to torsion: (a) rectangular section; (b) portion of box section with cantilever.

However, the contribution of the webs is accounted for through opposing shear forces in pairs of longitudinal beams and should not be accounted for again here. A formula suggested by Hambly (1991) halves the constant and removes the web term:

$$J_{cell} = \frac{2(hl)^2}{\left(\dfrac{l}{d_t} + \dfrac{l}{d_b}\right)}$$

$$\Rightarrow \quad J_{cell} = \frac{2h^2 l d_t d_b}{\left(d_t + d_b\right)} \tag{6.10}$$

EXAMPLE 6.9   SHEAR FLEXIBLE GRILLAGE MODEL OF A CELLULAR BRIDGE DECK

Figure 6.32 illustrates a two-span, three-cell bridge deck with edge cantilevers. There are 2 m thick solid diaphragms at the end and central supports. The deck is supported under each web at each support. A grillage model is required.

Figure 6.33 shows a convenient grillage mesh. Four longitudinal members are chosen, one at the centre of each web. The two edge members represent the portion of deck from the edge to halfway between the first and second webs (Figure 6.32). The two internal members represent the portion of deck from halfway between the first and second webs to the centre. Transverse grillage members are located at the ends and at the central support to represent the transverse

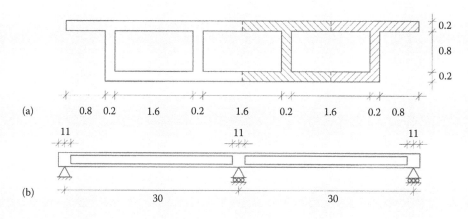

Figure 6.32 Cellular bridge of Example 6.9 (dimensions in m): (a) cross section; (b) longitudinal section.

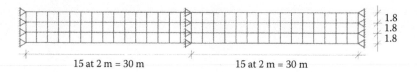

Figure 6.33 Plan view of grillage mesh.

diaphragms. Additional transverse members are placed at 2 m centres, giving a longitudinal-to-transverse member spacing ratio of 1:1.11.

The first step in determining the grillage member properties is to find the centroid of the deck. By summing moments of area about any point in the section, the centroid can be shown to be located at 0.65 m above the soffit. The second moments of area for the longitudinal members about this axis are then determined. For the edge longitudinal members

$$I_{edge} = 1.8(0.2)^3/12 + 1.8(0.2)(0.45)^2 + 1(0.2)^3/12 + 1(0.2)(0.55)^2$$
$$+ 0.2(0.8)^3/12 + 0.2(0.8)(0.05)^2$$

$$\Rightarrow \quad I_{edge} = 0.144 \text{ m}^4$$

For the internal longitudinal members

$$I_{int} = 1.8(0.2)^3/12 + 1.8(0.2)(0.45)^2 + 1.8(0.2)^3/12 + 1.8(0.2)(0.55)^2$$
$$+ 0.2(0.8)^3/12 + 0.2(0.8)(0.05)^2$$

$$\Rightarrow \quad I_{int} = 0.193 \text{ m}^4$$

For the transverse members, the centroid is at mid-depth, and the second moment of area per unit breadth i_{trans} is given by Equation 6.8:

$$i_{trans} = \frac{0.2^3}{12} + 0.2 \times 0.5^2 + \frac{0.2^3}{12} + 0.2 \times 0.5^2 = 0.101 \text{ m}^4/\text{m}$$

The breadth of the transverse members is 2 m, giving

$$I_{trans} = 0.202 \text{ m}^4$$

The torsion constant per cell is given by Equation 6.10:

$$J_{cell} = \frac{2h^2 l d_t d_b}{(d_t + d_b)} = \frac{2(1.0)^2(1.8)(0.2)(0.2)}{(0.2 + 0.2)} = 0.36 \text{ m}^4$$

This gives a torsion constant for the interior longitudinal members of 0.36 m⁴. The edge members only represent half a cell, and the contribution of the cantilever is added:

$$J_{edge} = \frac{0.36}{2} + \frac{0.8(0.2)^3}{3} = 0.182 \text{ m}^4$$

The torsion constant per unit breadth for the transverse members is taken to be equal to that of the longitudinal members:

$$j_{trans} = 0.36/1.8 = 0.2 \text{ m}^4/\text{m}$$

$$\Rightarrow \quad J_{trans} = 2(0.2) = 0.4 \text{ m}^4$$

The shear area per unit breadth of the transverse grillage members is given by Equation 6.7:

$$a_s = \left(\frac{E}{G}\right)\left(\frac{0.2^3(0.2^3 + 0.2^3)}{(1.8)(1.0)(0.2^3 + 0.2^3) + (1.8)^2(0.2)^3}\right) = 0.00234\left(\frac{E}{G}\right)$$

For concrete, a Poisson's ratio of 0.2 is assumed. Then, Equation 5.57 gives

$$G = \frac{E}{2(1+v)} = \frac{E}{2.4}$$

which results in a shear area of

$$a_s = 0.00561 \text{ m}^2/\text{m}$$

The breadth of the transverse members is 2 m, giving

$$A_{s\,trans} = 2 \times 0.00561 = 0.0112 \text{ m}^2$$

For the longitudinal members, the shear area is taken as the area of the web, a common approximation for I-sections, giving

$$A_{s\,long} = 1.0 \times 0.2 = 0.20 \text{ m}^2$$

The end and central diaphragm beams are 1.2 m deep by 2 m wide. The second moment of area of the grillage members representing these is therefore

$$I_{diaph} = \frac{2(1.2)^3}{12} = 0.288 \text{ m}^4$$

The torsion constant for the diaphragms is determined using Equation 6.2:

$$J_{diaph} = 2(1.2)^3\left[\frac{1}{3} - 0.21\frac{1.2}{2}\left(1 - \frac{1.2^4}{12(2)^4}\right)\right] = 0.721 \text{ m}^4$$

The shear area of the transverse diaphragm is taken as the actual shear area, as no significant transverse distortion is assumed to take place. For a rectangular section, the shear area can be shown to equal 83.3% of the actual area. Hence,

$$A_{s\,diaph} = 0.833(2 \times 1.2) = 2.0 \text{ m}^2$$

6.7 SKEW AND CURVED BRIDGE DECKS

Many bridge decks incorporate some degree of skew, and others are curved in plan. A grillage or FE model can be formulated for such decks based on the recommendations given in earlier sections, along with some additional considerations given here. Significant skew in bridge decks leads to a non-symmetrical distribution of reactions transversely between supports. Care is needed in modelling the support system in such cases, as any flexibility will cause a redistribution of reactions. The greatest reactions will tend to occur at obtuse corners in skew decks and the smallest reactions at acute corners. In highly skewed decks, uplift can occur at acute corners, which is generally to be avoided. Large reactions at obtuse corners lead to high shear forces, which can also be difficult to design for. A high degree of twisting, accompanied by large torsional moments (m_{xy}), is also associated with skew decks. As a result, in reinforced concrete, the Wood and Armer equations can dictate a requirement for top reinforcement near supports where hogging would not normally be expected.

6.7.1 Grillage modelling

A suitable grillage model of a skew deck will depend on the angle of skew, the span length and the width of the deck. An important consideration is to place the grillage members in the directions of principal strength. Figure 6.34a shows a long, narrow bridge deck with a high degree of skew, and Figure 6.34b shows a suitable grillage mesh. This deck will tend to span in the skew direction; thus the longitudinal grillage members are aligned in that direction. The transverse grillage members should generally be oriented perpendicular to the longitudinal members. An exception to this is in concrete decks where the transverse reinforcement is not perpendicular to the longitudinal reinforcement. In such cases, it is generally more appropriate to orientate the transverse members parallel to the transverse reinforcement, as illustrated in the alternative grillage layout of Figure 6.34c.

Figure 6.35a shows a short, wide bridge deck with a small angle of skew, and Figure 6.35b shows a suitable grillage layout. This deck will tend to span perpendicular to the supports rather than along the skew direction. Consequently, the longitudinal grillage members are orientated in this direction. Once again, the transverse grillage members are orientated perpendicular to the longitudinal members. Care should be taken with the edge grillage members, which generally will have to be orientated in the skew direction. If significant edge beams or stiffening is provided to the bridge deck, then this should be allowed for, when assigning the properties of the edge beams in the grillage. Bridge decks, which fall between

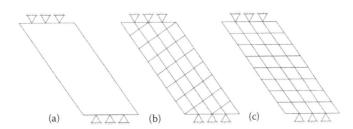

Figure 6.34 Long, narrow, highly skewed bridge deck: (a) plan view; (b) grillage layout; (c) alternative grillage layout.

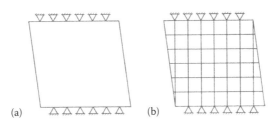

Figure 6.35 Short, wide bridge deck with small skew: (a) plan view; (b) grillage layout.

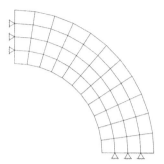

Figure 6.36 Grillage layout for curved bridge deck.

the extremes of Figures 6.34 and 6.35, will require a greater amount of judgement by the analyst in choosing a suitable grillage layout.

Curved decks pose no particular problem for grillage modelling. Some analysis programs will allow the use of curved beams, but straight beams are sufficiently accurate if the grillage mesh is fine enough. Figure 6.36 shows a suitable grillage mesh for a curved bridge deck. The longitudinal members, although straight, follow the curved layout closely due to the fineness of the mesh. The transverse members radiate from the centre of the curve. In this way, they are approximately perpendicular to the longitudinal members.

6.7.2 FE modelling

FE modelling of skew or curved decks should be carried out according to the recommendations for right (non-skewed) decks. Generally, no special consideration needs to be given to directions of strength as the elements are two-dimensional and will model the two-dimensional behaviour of the skew or curved slab. This is an advantage that the FE method has over the grillage method, especially for inexperienced users who might not be comfortable with the subtleties of grillage modelling. Skewed quadrilateral elements, as illustrated in Figure 6.37a, can give results that are just as accurate as those for rectangular elements, and they are very easy to implement. However, highly skewed quadrilaterals may result in round-off errors due to calculations involving small angles. In such cases, rectangular elements, with some triangular elements near the ends, may be used, as illustrated in Figure 6.37b.

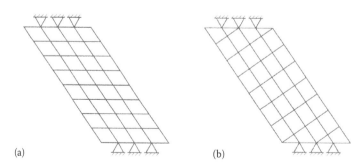

(a)   (b)

Figure 6.37 Alternative FE meshes: (a) skewed quadrilateral elements; (b) alternative for high skew, with rectangular and triangular elements.

Chapter 7

Three-dimensional modelling of bridge decks

7.1 INTRODUCTION

In Chapters 5 and 6, the analysis of bridge decks using planar models is discussed. In this chapter, more sophisticated three-dimensional models are described. Bridge decks with wide edge cantilevers are an example of the kind of problem that may require a three-dimensional analysis. Decks with edge cantilevers are considered in Chapter 6, but it is stipulated that only those with short cantilevers should be analysed using planar models. Three-dimensional models require much more computing time than planar models, and considerable post-processing is generally required to make sense of the results. However, they can be used to great effect to determine the stresses in bridges with more complex geometries.

7.2 SHEAR LAG AND EFFECTIVE FLANGE WIDTH

When a bridge deck flexes, longitudinal bending stresses occur. These are distributed transversely from one part of the deck to adjacent parts by interface shear stresses. Thus, when the bending moment in a flanged beam varies from one point to another, interface stresses are generated, as illustrated in Figure 7.1. When flanges or cantilevers are wide and slender, the edges do not receive the same amount of axial stress as those near the centre of the bridge. This phenomenon is known as 'shear lag' as it is associated with interface shear and is characterised by the lagging behind of axial stresses at the edges of cantilevers. The extent of the reduction of stress is dependent on both the geometric shape of the bridge deck and the nature of the applied loading.

Figure 7.2a shows a bridge deck with the edge cantilevers separated from the main part of the deck. If a load were applied to the deck in this condition, each part would bend about its own centroid, independently of the rest. In this condition, the bridge deck has a discontinuous neutral axis as indicated in the figure. If the bridge deck is now rejoined, a common centroid can be found (Figure 7.2b), and the entire bridge is often assumed to bend about this. As the rejoined bridge bends, the remote edges of the cantilevers, due to shear lag, do not experience the same amount of axial stress as the main part of the deck, as can be seen in Figure 7.2c. The effect of bending is not felt to the same extent in the edges of the cantilevers as it is elsewhere. This is because the edges of the cantilevers tend to bend about their own centroidal axes. Obviously, they are not free to do this, but this tendency causes the overall bridge deck centroid to move towards the centroid of the cantilevers at the edges. Such a non-uniform neutral axis is illustrated in Figure 7.2d.

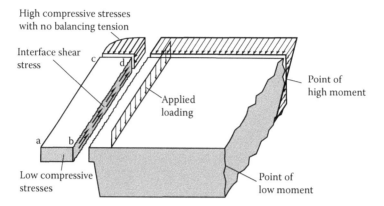

High compressive stresses
with no balancing tension

Interface shear
stress

Applied
loading

Point of
high moment

Low compressive
stresses

Point of
low moment

Figure 7.1 Interface shear stresses in flanged beam subject to bending.

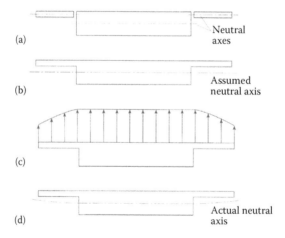

(a)

Neutral
axes

(b)

Assumed
neutral axis

(c)

(d)

Actual neutral
axis

Figure 7.2 Transverse variation in neutral axis: (a) if cantilevers and main deck were free to act independently; (b) commonly assumed straight axis; (c) variation in longitudinal stress at top of deck;
(d) effective neutral axis location.

There is a strong link between shear lag and neutral axis location. It could be said that
the variation in the axis location in a bridge deck is caused by shear lag or that shear lag is
caused by the tendency of each part of the bridge deck to bend about its own axis. A three-
dimensional analysis can automatically account for shear lag as it allows for variations in
neutral axis location directly.

7.2.1 Effective flange width

In the design of bridge decks, a two-dimensional analysis, as described in Chapter 6, is
often used which does not take account of shear lag. It is possible to overcome this problem
by assuming an 'effective flange width' for the edge cantilevers, as illustrated in Figure 7.3.
The method uses a notional width of cantilever in the grillage or finite element (FE) model,
which has a uniform stress distribution whose force is equal in magnitude to the actual
force. Hence, a two-dimensional model with an effective flange width, analysed with no
allowance for shear lag, can be used to analyse the bridge.

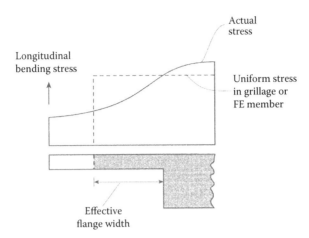

Figure 7.3 Actual and calculated distributions of longitudinal bending stress at top of flanged deck.

The correct effective flange width to be used for the cantilever is largely dependent on the ratio of the actual cantilever width to the length between points of zero moment (points of contraflexure), as it is from these points that longitudinal stresses begin to spread out into the cantilevers. The effective flange width is also dependent on the form of the applied loading. Hambly (1991) presents a chart for the determination of effective flange widths for beams subjected to distributed and concentrated loads. The chart, reproduced here as Figure 7.4, relates the ratio of effective flange width, b_e, and actual flange width, b, to the ratio of actual flange width, b, and length between points of contraflexure, L. Also shown in Figure 7.4 are the popular approximations for this relationship:

$b_e = L/6$ (for uniform loading)

and

$b_e = L/10$ (for concentrated loading)

which can be seen to be reasonably accurate for relatively wide flanges.

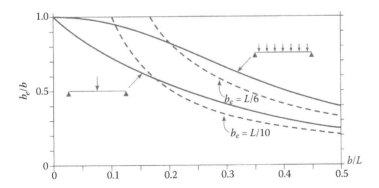

Figure 7.4 Effective flange width for different loadings (solid lines) and common approximations (dashed lines).

EXAMPLE 7.1   EFFECTIVE FLANGE WIDTH

Figure 7.5a shows the cross section of a bridge deck with edge cantilevers. The cantilevers are 2.4 m wide, and the deck has a single simply supported span of 20 m. Find the effective flange widths.

As the span is simply supported, the length between points of contraflexure, L, is equal to the span length in this case. Hence, the ratio of the cantilever width to this length, b/L, is 2.4/20 = 0.12. From Figure 7.4, the ratios of b_e/b are 0.93 and 0.67 for the uniformly distributed and point load cases, respectively. This results in effective flange widths of 2.23 and 1.61 m, respectively. Figure 7.5b shows the effective flange width for one of these load cases. A constant stress is assumed in the modelled portion of the cantilever, and that part of it outside the effective flange width is ignored.

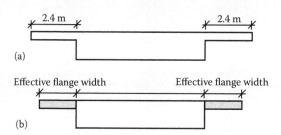

Figure 7.5 Cross section of bridge deck of Example 7.1: (a) showing actual cantilever widths; (b) showing effective flange widths.

This example highlights the limitations of the effective flange width method as the nature of the loading causes a substantial variation in the effective flange width.

7.3 THREE-DIMENSIONAL ANALYSIS USING BRICK ELEMENTS

The use of two-dimensional analysis methods with effective flange widths is approximate at best and does not address the issue of upstands, which are often provided at the edges of bridge cantilevers. Bridges with significant shear lag are just one example where three-dimensional models are necessary to achieve an accurate representation of the behaviour of the structure.

The most sophisticated three-dimensional FE models use solid 'brick'-type elements. Figure 7.6 shows such a model of a portion of bridge deck with edge cantilevers. The benefit of this type of model is that it can be used to describe the geometry of highly complex bridge decks very accurately. The inclusion of voids, a cellular structure or transverse diaphragms poses no particular problems, provided the mesh is sufficiently dense. The model automatically allows for any variations in the location of the neutral axis and hence allows for shear lag in edge cantilevers.

7.3.1 Interpretation of results of brick models

One of the perceived shortcomings of three-dimensional brick models among designers is the great quantity of results data created. As discussed in Chapter 5, there are six components of stress at a point. It follows that the brick model generates six stress components for

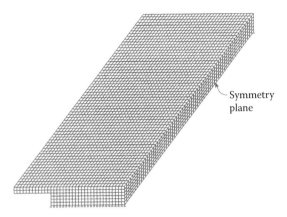

Figure 7.6 Portion of bridge deck modelled with solid brick elements. (Note that only two layers of elements are shown in the cantilever region. This is a very crude mesh if the elements only have translational degrees of freedom. For accurate results, these elements would also need rotational degrees of freedom.)

each brick element. For designers, it is more useful to combine element stress components into the moments and forces per unit breadth commonly used in design.

Figure 7.7 illustrates a single column of brick elements through the depth of a slab. For the element highlighted, the six components of stress are the three axial components, σ_x, σ_y and σ_z, and the three shear components, $\tau_{xy} = \tau_{yx}$, $\tau_{xz} = \tau_{zx}$ and $\tau_{yz} = \tau_{zy}$. The vertical (Z direction) bearing stress, σ_z, is generally not of interest.

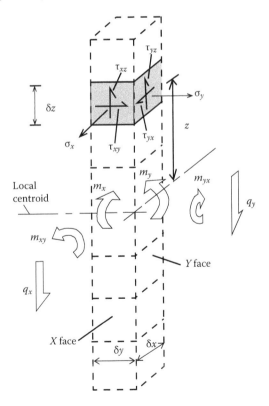

Figure 7.7 Column of brick elements through depth of slab.

The axial stress on the X face of the column generates bending moment and sometimes also axial force. This stress, σ_x, generates a moment at the local centroid of σ_x $(\delta y \times \delta z)z$. Hence, the contribution of this element to moment per unit breadth on the X face is

$$m_x = -\sigma_x(\delta z)z$$

Summing or integrating through the depth of the slab then gives

$$m_x = -\int \sigma_x z \, dz \approx -\sum \sigma_{xi} z_i \delta z \tag{7.1}$$

where i is the element number in the column of elements. (This is Equation 5.30 in Chapter 5.)
The stresses can also be summed to get the axial force per unit width, f_x:

$$f_x = \int \sigma_x \, dz \approx \sum \sigma_{xi} \delta z \tag{7.2}$$

It is common practice to assume that there are no longitudinal axial forces in slabs, but where shear lag exists, there will be axial forces in some parts of the slab that are balanced by axial forces in other parts.
Similar expressions can be derived for the Y face of the column of elements:

$$m_y = -\int \sigma_y z \, dz \approx -\sum \sigma_{yi} z_i \delta z \tag{7.3}$$

$$f_y = \int \sigma_y \, dz \approx \sum \sigma_{yi} \delta z \tag{7.4}$$

where f_y is the axial force per unit breadth in the Y direction. The twisting moments are found by summing the contributions of the τ_{xy} stresses, noting that τ_{xy} refers to shear stress on the X face and in the Y direction. Hence,

$$m_{xy} = -\int \tau_{xy} z \, dz \approx -\sum \tau_{xyi} z_i \delta z \tag{7.5}$$

Finally, the vertical components of shear stress are summed to find the shear intensities. On the X face, for example, the shear force per unit breadth is

$$q_x = -\int \tau_{xz} \, dz \approx -\sum \tau_{xzi} \delta z \tag{7.6}$$

Similarly, on the Y face, the shear intensity is

$$q_y = -\int \tau_{yz} \, dz \approx -\sum \tau_{yzi} \delta z \tag{7.7}$$

The evaluation of these moments and forces per unit breadth is illustrated in the following examples.

EXAMPLE 7.2   THREE-DIMENSIONAL BRICK MODEL OF SOLID SLAB

The slab bridge of Figure 7.8a is made up of inverted-T pre-tensioned prestressed concrete beams and in situ concrete. The prestressed concrete has cylinder strength of f_{ck} = 45 N/mm², while the in situ concrete has cylinder strength of f_{ck} = 35 N/mm². Each beam is supported at each end. Carry out a three-dimensional brick FE analysis to determine (1) the distribution of longitudinal moments at the centre and (2) the shear intensities at one slab depth from the supports due to self-weight.

As the in situ concrete is fully bonded to the prestressed beams, the slab can be treated as solid. While there are two grades of concrete, the difference in their stiffnesses is small: the Eurocode, EN 1992-1-1 specifies mean values for Young's modulus of 36,000 N/mm² for f_{ck} = 45 N/mm² and 34,000 for f_{ck} = 35 N/mm². In any event, it is only the relative values of the Young's moduli that influence the distribution of stresses/moments and these are the same, on average, in the two directions. Therefore, one modulus is chosen, E_{cm} = 35,000 N/mm², and the slab is treated as a solid rectangular block.

A 0.1 × 0.1 × 0.1 m³ brick element is chosen with six degrees of freedom per node. This gives only eight elements through the depth, but it should be borne in mind that the elements have rotational degrees of freedom, so each element has the ability to bend as well as to expand and contract. The FE mesh is illustrated in Figure 7.8b. The model is analysed for the effects of self-weight, assuming a concrete density of 25 kN/m³.

1. Longitudinal moments

    The longitudinal axial stresses near the edge, at mid-span, are given in Table 7.1. Summing transversely for each 1 m strip of slab gives the axial forces per metre of Table 7.2. Hence, for example, in the strip of slab centred 0.5 m from the edge (y = 0.5 m), the axial force in the first layer of elements is (−5324 −5330− ... −5296) × 0.1 × 0.1 = −532 kN/m. Summing moments about the mid-depth gives the moments per metre presented in Table 7.3. For example, for y = 0.5 m, m_x = −[−532 × 0.35 − 379 × 0.25 − ... + 532 × (−0.35)] = 638 kNm/m. Note that the average moment across the slab is 627 kNm/m, which is approximately 2% short of the theoretical mid-span moment of $wl^2/8$ = (25 × 0.8)(16)²/8 = 640 kNm/m. This is

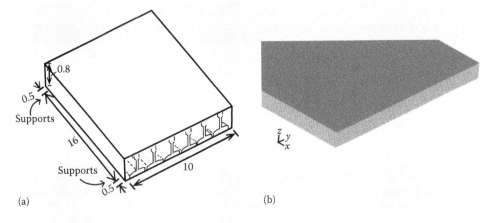

(a)   (b)

Figure 7.8 Bridge slab of Example 7.2: (a) bridge (dimensions in m); (b) FE mesh.

Table 7.1 Axial stresses, σ_{xi}, at the centres of the brick elements (kN/m²)

Distance from mid-depth, z (m)	Transverse distance from edge, y (m)										
	0.05	0.15	0.25	0.35	0.45	0.55	0.65	0.75	0.85	0.95	1.05
0.35	−5324	−5330	−5331	−5329	−5325	−5320	−5314	−5308	−5302	−5296	−5290 ...
0.25	−3796	−3801	−3803	−3803	−3801	−3797	−3794	−3789	−3785	−3781	−3776 ...
0.15	−2275	−2279	−2280	−2280	−2279	−2278	−2275	−2273	−2270	−2268	−2265 ...
0.05	−758	−760	−760	−760	−760	−760	−759	−758	−757	−756	−755 ...
−0.05	757	758	759	759	758	758	757	757	756	755	754 ...
−0.15	2273	2277	2279	2279	2278	2276	2274	2272	2269	2267	2264 ...
−0.25	3794	3800	3802	3801	3799	3796	3792	3788	3784	3779	3775 ...
−0.35	5322	5328	5329	5327	5323	5318	5313	5307	5301	5295	5288 ...

Table 7.2 Axial forces per metre in each layer at the centres of 1 m strips of slab (kN/m)

Distance from mid-depth, z (m)	Transverse distance from edge, y (m)									
	0.5	1.5	2.5	3.5	4.5	5.5	6.5	7.5	8.5	9.5
0.35	−532	−526	−522	−519	−517	−517	−519	−522	−526	−532
0.25	−379	−376	−372	−370	−369	−369	−370	−372	−376	−379
0.15	−228	−225	−223	−222	−221	−221	−222	−223	−225	−228
0.05	−76	−75	−74	−74	−74	−74	−74	−74	−75	−76
−0.05	76	75	74	74	74	74	74	74	75	76
−0.15	227	225	223	222	221	221	222	223	225	227
−0.25	379	376	372	370	369	369	370	372	376	379
−0.35	532	526	522	519	517	517	519	522	526	532

Table 7.3 Moments per metre at the centres of 1 m strips of slab

	Transverse distance from edge, y (m)									
	0.5	1.5	2.5	3.5	4.5	5.5	6.5	7.5	8.5	9.5
Moment, m_x	638	631	626	622	620	620	622	626	631	638

largely due to the number of elements through the depth – more elements would increase the accuracy.

2. Shear forces

Some of the vertical shear stresses on the X face are given in Table 7.4a and b for the two locations nearest to $x = 0.8$ m at which results are available (element centres). Interpolating gives the shear stresses at $x = 0.8$ presented in Table 7.4c.

Summing forces vertically for each column of elements and transversely for each 1 m strip of slab gives the shear forces per metre of Table 7.5. Hence, for example, for $y = 0.5$ m, the shear intensity is, $-[(-181 - 443 - \ldots - 173) \times 0.1 + (-125 - 336 - \ldots - 118) \times 0.1 + \ldots + (-63 - 164 - \ldots -58) \times 0.1] \times 0.1 = 205$ kN/m.

Table 7.4 Vertical shear stresses, τ_{xzi}, at the centres of the brick elements (kN/m²)

Distance from mid-depth, z (m)	Transverse distance from edge, y (m)											
	0.05	0.15	0.25	0.35	0.45	0.55	0.65	0.75	0.85	0.95	1.05	...
(a) At x = 0.75												
0.35	−182	−126	−104	−90	−81	−75	−71	−68	−65	−64	−62	...
0.25	−445	−339	−278	−240	−215	−198	−186	−178	−172	−167	−164	...
0.15	−598	−477	−395	−340	−304	−279	−262	−249	−241	−234	−229	...
0.05	−666	−540	−450	−388	−345	−316	−296	−282	−272	−265	−259	...
−0.05	−659	−534	−444	−382	−339	−311	−291	−278	−268	−261	−255	...
−0.15	−580	−460	−379	−325	−289	−265	−249	−237	−229	−224	−219	...
−0.25	−425	−321	−260	−222	−198	−182	−172	−164	−159	−155	−153	...
−0.35	−172	−117	−95	−81	−73	−67	−64	−61	−59	−58	−57	...
(b) At x = 0.85												
0.35	−180	−124	−101	−87	−78	−72	−68	−65	−63	−61	−60	...
0.25	−440	−334	−271	−233	−208	−191	−180	−172	−166	−162	−159	...
0.15	−592	−470	−388	−332	−296	−271	−254	−242	−234	−228	−223	...
0.05	−662	−535	−444	−381	−338	−309	−289	−275	−266	−259	−253	...
−0.05	−657	−530	−439	−377	−334	−305	−286	−272	−263	−256	−251	...
−0.15	−580	−460	−377	−323	−286	−262	−246	−234	−226	−221	−217	...
−0.25	−427	−322	−260	−222	−198	−182	−171	−163	−158	−154	−152	...
−0.35	−174	−118	−96	−82	−73	−68	−64	−61	−59	−58	−57	...
(c) At x = 0.8 m												
0.35	−181	−125	−102	−88	−79	−73	−69	−66	−64	−63	−61	...
0.25	−443	−336	−275	−236	−211	−195	−183	−175	−169	−164	−161	...
0.15	−595	−474	−391	−336	−300	−275	−258	−246	−237	−231	−226	...
0.05	−664	−537	−447	−384	−341	−313	−293	−279	−269	−262	−256	...
−0.05	−658	−532	−442	−379	−337	−308	−289	−275	−265	−258	−253	...
−0.15	−580	−460	−378	−324	−288	−263	−247	−236	−228	−222	−218	...
−0.25	−426	−321	−260	−222	−198	−182	−171	−164	−159	−155	−152	...
−0.35	−173	−118	−95	−82	−73	−67	−64	−61	−59	−58	−57	...

Table 7.5 Shear forces per metre at the centres of 1 m strips of slab (kN/m)

	Transverse distance, y (m)									
	0.5	1.5	2.5	3.5	4.5	5.5	6.5	7.5	8.5	9.5
Shear intensity, q_x	205	132	127	128	129	129	128	127	132	205

It can be seen that there is a concentration of shear intensity near the edge of the slab as noted by Timoshenko and Woinowsky-Krieger (1970). For this example, the shear in the edge 1 m strip is 42% greater than the average across the width. It should be noted that this edge effect is not generally found in a planar FE or grillage analysis.

EXAMPLE 7.3   THREE-DIMENSIONAL BRICK
MODEL OF SLAB WITH CANTILEVER

The bridge deck illustrated in Figure 7.9 is constructed of in situ concrete and includes 1.5 m cantilevers at the edges. It is post-tensioned with nine parabolically profiled tendons, at 1 m centres, that vary from mid-depth over the supports to 0.1 m from the soffit (bottom surface) at the centre. The force in each tendon is 2650 kN, and no allowance needs to be made for friction losses. The bridge is supported on two pot bearings at each end, as shown. A three-dimensional brick element FE analysis is required to determine (a) the axial stress and moment at mid-span near the edges due to self-weight and prestress and (b) the shear intensity due to self-weight near the supports.

The dimensions of the FE mesh are generally the same as those of the actual structure (Figure 7.10a). As in the previous example, a brick element size of 0.1 × 0.1 × 0.1 m³ is adopted, and

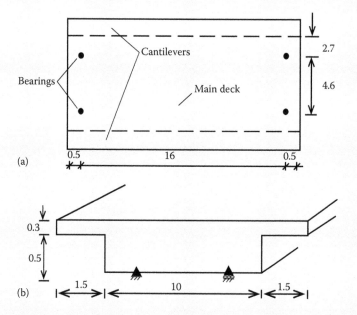

Figure 7.9 Bridge slab of Example 7.3: (a) plan; (b) cross section.

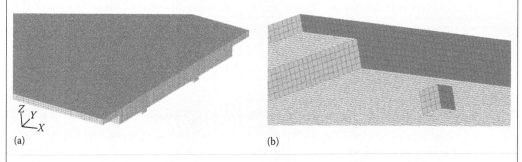

Figure 7.10 Brick FE mesh for slab of Example 7.3: (a) overall view from above; (b) view from below showing detail at support.

each element has six degrees of freedom per node. (The cantilever will be only three elements deep; thus, the rotational degrees of freedom are necessary to give a good model.) Pot bearings provide continuous support over an area of approximately 400×400 mm² to the deck soffit. To reproduce this, a block of elements is extended downwards from the bridge soffit sufficiently far to allow the point force generated by the model support to be dispersed across the area (Figure 7.10b).

The self-weight loading is generated by specifying a material density of 25 kN/m³. The pre-stress loading is generated using the recommendations of Section 2.8 (see Example 2.5). The tendons are at mid-depth over the supports, that is, 0.4 m from the soffit and 0.1 above the soffit at mid-span. Hence, the end eccentricities are equal, and the sag, s, is 0.3. From Figure 2.22 and Equation 2.19, each tendon applies a uniformly distributed upwards loading of

$$w = \frac{8Ps}{l^2} = \frac{8(2650)(0.3)}{(16)^2} = 25 \text{ kN/m}$$

The equivalent moments due to prestress at the ends are less obvious. The extent of shear lag is not known; thus, the depth to the axis of bending near the edges can only be estimated. However, with a brick FE analysis, this is not a problem; the equivalent force due to prestress is simply applied at the depth of the tendon, and the corresponding equivalent moment does not need to be calculated.

To get the correct distribution of shear, the vertical and horizontal components of the equiv-alent force are each applied. The horizontal component is $P\cos\theta \approx P$, where P is the prestress force and θ is the angle of the tendon at the supports. At the left support, this is defined by Equation 2.16:

$$\sin\theta_A \approx \frac{e_B - e_A - 4s}{l}$$

This equation is in fact only valid when shear lag is negligible and can be more generally written as

$$\sin\theta_A \approx \frac{h_B - h_A - 4s}{l} \tag{7.8}$$

where h_A and h_B are the heights of the tendon at the left and right supports, respectively, above a specified datum. For this example, the heights are equal; thus, the vertical component of pre-stress is

$$P\sin\theta_A \approx \frac{-4Ps}{l} = \frac{-4(2650)(0.3)}{16} = -199 \text{ kN}$$

In summary, the equivalent loading due to each tendon consists of

- A uniformly distributed upwards loading of $w = 25$ kN/m
- A horizontal point force at each end of $P = 2650$ kN, 0.4 m above the soffit (as the force is not applied at the centroid, this will generate an equivalent moment)
- A vertical downwards point force at each end of 199 kN

Table 7.6 Axial stresses, σ_{xt}, at the centres of the brick elements (kN/m²)

Distance from mid-depth, z (m)	Transverse distance from edge, y (m)																				
	0.05	0.15	0.25	0.35	0.45	0.55	0.65	0.75	0.85	0.95	1.05	1.15	1.25	1.35	1.45	1.55	1.65	1.75	1.85	1.95	...
(a) Self-weight																					
0.35	−4240	−4249	−4259	−4270	−4283	−4298	−4314	−4332	−4351	−4372	−4395	−4419	−4445	−4471	−4502	−4537	−4571	−4603	−4631	−4654	...
0.25	−2771	−2731	−2742	−2756	−2772	−2791	−2811	−2834	−2859	−2886	−2915	−2946	−2980	−3020	−3055	−3088	−3112	−3131	−3151	−3170	...
0.15	−1201	−1212	−1225	−1242	−1261	−1283	−1308	−1336	−1366	−1399	−1434	−1473	−1514	−1558	−1630	−1653	−1652	−1661	−1674	−1690	...
0.05																−205	−184	−189	−198	−210	...
−0.05																1294	1290	1289	1281	1270	...
−0.15																2780	2778	2771	2763	2752	...
−0.25																4272	4267	4260	4249	4237	...
−0.35																5768	5763	5753	5740	5726	...
(b) Prestress																					
0.35	2299	2312	2325	2338	2351	2366	2381	2397	2415	2433	2453	2474	2495	2518	2537	2554	2566	2576	2585	2593	...
0.25	639	652	665	680	694	710	726	744	762	781	801	822	844	865	891	914	936	953	964	974	...
0.15	−1022	−1009	−994	−979	−962	−946	−928	−910	−891	−872	−851	−830	−807	−782	−764	−718	−684	−665	−653	−643	...
0.05																−2321	−2293	−2278	−2267	−2258	...
−0.05																−3900	−3898	−3887	−3879	−3871	...
−0.15																−5504	−5502	−5497	−5491	−5485	...
−0.25																−7113	−7113	−7109	−7104	−7099	...
−0.35																−8731	−8729	−8726	−8721	−8717	...

a. Axial effects due to self-weight and prestress

The axial stresses near the edge at mid-span, σ_{xi}, are found from the FE analysis. The results for the self-weight and prestress load cases are given in Table 7.6. It can be seen that the neutral axis is below the cantilever for the self-weight load case (the axial stresses are all compressive).

Clearly, the axial stresses due to a combination of prestress and self-weight can be combined directly by adding the corresponding stresses in Table 7.6a and b. Alternatively, the axial stresses can be post-processed to find the corresponding bending moments and axial forces per metre. Following the same approach adopted for Example 7.2, the axial forces and moments due to self-weight are found and presented in Table 7.7. It is important to note that these moments are calculated at mid-depth of each section, and the mid-depth is different in the cantilever and the main deck. If, for example, 2.5 m of bridge at the edge were treated as a single unit, the moment would need to be calculated by combining the relevant moments per metre in Table 7.7 with the contribution of the axial forces to a shared axis of bending (see Figure 7.11).

Bearing in mind that the edge strip in the cantilever is 0.5 m wide, the total moment is (30/2 + 30 + 622) = 667 kNm. However, there is a net axial compression in this 2.5 m edge section of 866 kN. Summing moments and forces shows that this is equivalent to an axial force of 866 kN at a point 0.741 m above the top of the bridge – the applied moment is strongly influenced by the point about which it is calculated. That point at

Table 7.7 Axial forces per metre, f_x, and moments per metre, m_x, due to self-weight at the centres of strips of slab

| | Transverse distance from edge, y (m) | | | | | | | | | | | | | |
	0.25	1.0	2.0	3.0	4.0	5.0	6.0	7.0	8.0	9.0	10.0	11.0	12.0	12.75
Force, f_x (kN/m)	−823	−873	419	314	231	175	146	146	175	231	314	419	−873	−823
Moment, m_x (kNm/m)	30	30	622	624	625	626	627	627	626	625	624	622	30	30

Note: All strips are 1 m wide except the cantilever edge strips, which are 0.5 m.

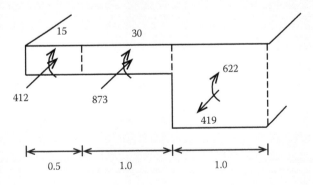

Figure 7.11 Equivalent moments and forces in edge section.

Table 7.8 Vertical shear stresses (kN/m²) due to self-weight at the centres of brick elements in the region of a support, τ_{xzi}

Distance from centre, z (m)		Transverse distance from centre of bearing (m)												
		-0.55	-0.45	-0.35	-0.25	-0.15	-0.05	0.05	0.15	0.25	0.35	0.45	0.55	
At edge of bearing (x = 0.5 + 0.4/2 ≈ 0.75 m)														
0.35	...	-86	-99	-112	-125	-134	-139	-139	-134	-125	-112	-99	-86	...
0.25	...	-222	-255	-288	-319	-343	-356	-356	-343	-319	-289	-255	-223	...
0.15	...	-311	-359	-411	-459	-497	-517	-517	-497	-460	-412	-360	-312	...
0.05	...	-359	-423	-496	-567	-623	-654	-654	-624	-568	-498	-425	-361	...
-0.05	...	-363	-446	-549	-655	-741	-787	-788	-743	-659	-552	-449	-366	...
-0.15	...	-317	-413	-563	-728	-856	-921	-924	-864	-737	-570	-417	-319	...
-0.25	...	-216	-291	-486	-806	-1040	-1097	-1107	-1065	-831	-498	-294	-218	...
-0.35	...	-77	-96	-172	-709	-1240	-1270	-1292	-1303	-755	-178	-98	-78	...
Half a deck depth from edge of bearing (x = 1.15 m)														
0.35	...	-99	-110	-121	-131	-138	-142	-142	-139	-131	-122	-111	-100	...
0.25	...	-256	-284	-311	-335	-354	-364	-365	-355	-338	-314	-287	-260	...
0.15	...	-356	-394	-431	-464	-490	-504	-505	-492	-468	-435	-399	-362	...
0.05	...	-402	-444	-485	-524	-553	-570	-571	-556	-528	-491	-449	-408	...
-0.05	...	-395	-435	-476	-513	-541	-557	-559	-544	-518	-482	-441	-401	...
-0.15	...	-337	-370	-402	-432	-454	-467	-468	-458	-436	-408	-375	-343	...
-0.25	...	-232	-252	-272	-288	-301	-308	-309	-303	-292	-276	-256	-236	...
-0.35	...	-86	-92	-98	-102	-106	-107	-108	-107	-104	-99	-94	-87	...
Full deck depth from edge of bearing (x = 1.55 m)														
0.35	...	-81	-85	-89	-92	-94	-96	-96	-95	-93	-90	-87	-83	...
0.25	...	-213	-224	-234	-242	-248	-251	-252	-249	-244	-237	-228	-218	...
0.15	...	-301	-315	-328	-340	-348	-352	-353	-349	-342	-332	-320	-306	...
0.05	...	-343	-359	-373	-386	-394	-399	-400	-396	-389	-378	-365	-349	...
-0.05	...	-341	-356	-370	-382	-390	-395	-395	-392	-385	-375	-362	-347	...
-0.15	...	-295	-308	-320	-329	-336	-340	-341	-338	-332	-324	-313	-301	...
-0.25	...	-208	-217	-225	-231	-236	-238	-239	-237	-233	-228	-220	-212	...
-0.35	...	-79	-82	-85	-87	-89	-90	-90	-89	-88	-86	-83	-80	...

which the applied moment is calculated should clearly match the point about which the moment *capacity* is calculated.

b. Shear intensity due to self-weight

In Chapter 5, it is pointed out that neither grillage nor FE plate models are suited to the calculation of shear intensities near point supports. In both cases, the calculated shear intensity tends towards infinity near the support as the mesh density is increased. Fortunately, a three-dimensional brick model can accurately model the dispersion of vertical shear stresses that occurs through the depth of the deck. For this example, the shear stresses due to self-weight on the longitudinal face in the region of a support, τ_{xzi}, are presented in Table 7.8.

The forces corresponding to these stresses are summed through the depth of the deck, and the resulting shear intensities are presented in Table 7.9. It can be seen that shear intensities are relatively high in the region of the support, but they are not approaching infinity (as a plate FE model might suggest). It should also be noted that shear enhancement applies in this region.

Table 7.9 Shear intensities (kN/m) due to self-weight in the region of a support

		Transverse distance from centre of support (m)											
	...	−0.55	−0.45	−0.35	−0.25	−0.15	−0.05	0.05	0.15	0.25	0.35	0.45	0.55 ...
Edge of bearing	...	195	238	308	437	547	574	578	557	445	311	240	196 ...
Half depth from edge	...	216	238	260	279	294	302	303	295	281	263	241	220 ...
Full depth from edge	...	186	195	202	209	214	216	217	215	211	205	198	190 ...

7.4 UPSTAND GRILLAGE MODELLING

In Chapter 6, grillage modelling is applied to bridge decks, including those with edge cantilevers. That type of analysis is referred to as planar grillage as all of the grillage members are located in one plane. It is suitable for bridge decks where the neutral axis remains substantially straight across the deck and is coincident with the centroidal axis of the bridge. Hence, it works well for solid slab bridges with short cantilevers. Planar grillage can also be used for beam-and-slab bridges – while each part of the bridge bends about its own axis, the differences in the neutral axis depths are not generally great.

When there is significant variation in neutral axis depth, a three-dimensional technique, such as upstand grillage modelling, is required. The upstand grillage analogy is a direct extension of the planar grillage analogy but involves the modelling of each part of the bridge deck as a separate plane grillage located at the centroid of the portion of bridge deck which it represents. The plane grillage meshes are then connected using rigid vertical members. Figure 7.12 shows an upstand grillage model for a bridge deck with edge cantilevers. In this, the edge cantilevers are modelled with grillage members that are located at the centroid of the cantilevers, while the main part of the deck is modelled with grillage members located at the centroid of that part.

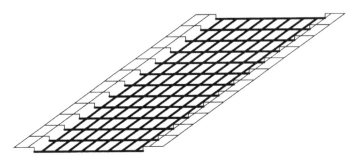

Figure 7.12 Upstand grillage model.

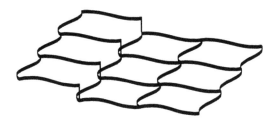

Figure 7.13 In-plane distortion of members in upstand grillage model.

The properties of each part of the deck are determined relative to their own centroids. Consequently, there is no need to make an assumption as to the location of the overall bridge neutral axis. There is also no need to assume an effective flange width to allow for shear lag effects. As the model is three-dimensional, it should automatically determine the location of the neutral axis, be it straight or varying, for each load case considered. Consequently, shear lag, where it exists, should be accounted for automatically.

Although the upstand grillage seems to be a relatively simple and powerful model, difficulties arise when in-plane effects are considered. Unlike the plane grillage, the three-dimensional nature of the model causes local in-plane distortions in the grillage members, as illustrated in Figure 7.13, which are clearly inconsistent with the behaviour of the bridge deck. Such behaviour in the model could be addressed in one of two ways. The members could be given very large in-plane second moments of area, or the nodes at the ends of the members could be restrained against in-plane rotation. However, such an approach is not recommended, given that upstand FE models are widely available and are not susceptible to such local distortions.

7.5 UPSTAND FINITE ELEMENT MODELLING

Upstand FE modelling is an extension of plane FE modelling in the same way that upstand grillage modelling is an extension of plane grillage modelling. The upstand FE model consists of a number of planes of plate FEs connected together by rigid vertical members. Vertical members can be short beams or plate elements. Figure 7.14 shows an upstand FE model for a bridge deck with edge cantilevers. The cantilevers are idealised with FEs located at the level of the centroids of the actual cantilevers, while the main part of the deck is idealised using elements located at the centroid of that part. The FE meshes on each plane are connected by rigid vertical beam members.

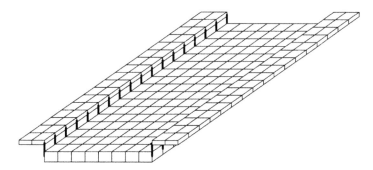

Figure 7.14 Upstand FE model.

In a series of tests, the authors have found the upstand FE method to be very suitable for modelling bridge decks with wide edge cantilevers. It benefits from being three-dimensional while being relatively simple to use. Most significantly, it does not suffer from the problems of modelling in-plane behaviour associated with upstand grillages. This is largely due to the well-proven ability of FEs to model in-plane behaviour.

Figure 7.15a shows the cross section of a 24.8 m single-span bridge deck with wide edge cantilevers (OBrien and Keogh 1998). This bridge was analysed by Keogh and OBrien (1996) under the action of a constant longitudinal bending moment using a number of models, including a full three-dimensional brick model similar to that shown in Figure 7.6. Figure 7.15b shows an exaggerated plan view of the deflected shape of the three-dimensional brick FE model (only one half of the model is shown as it is symmetrical). The in-plane distortion seen

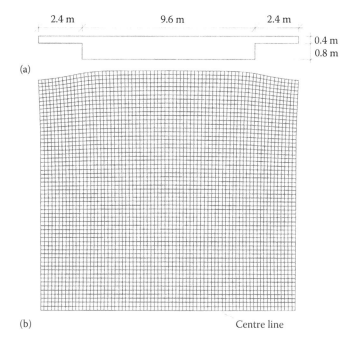

Figure 7.15 In-plane deformation in cantilevers of deck: (a) cross section; (b) plan view of deflected shape (only half is shown).

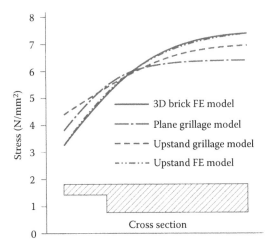

Figure 7.16 Calculated longitudinal bending stresses at 1/8-span on top surface of deck.

at the end of the cantilevers is made up of both in-plane shear distortion and in-plane bending. Figure 7.16 shows the longitudinal bending stress predicted along the top of this bridge deck at 1/8 span by the various models. Only half of the width is shown, and the cross section is included for reference.

Assuming the three-dimensional brick model to be the most accurate, the benefits of the upstand FE model can be seen as it very accurately tracks the stresses in the brick model. However, this is not the case for plane grillage or upstand grillage. Similarly good results for the upstand FE model were found at mid-span and for all other cases considered.

EXAMPLE 7.4   UPSTAND FINITE ELEMENT MODEL

Figure 7.17a shows the cross section of a bridge deck with wide edge cantilevers. The deck is continuous over two spans of 24.8 m and is supported along the entire width of the main part of the deck at each support location. An upstand FE model is required.

Figure 7.17b shows a three-dimensional view of a suitable upstand FE mesh. All of the elements are 1.2 m wide and 1.24 m long (in the span direction). The elements representing the edge cantilevers are located at the centroid of the cantilevers, which is 0.2 m below the top of the bridge deck. Those representing the main part of the deck are located at the centroid of that part, which is 0.6 m from the top of the deck. This results in vertical members with a length of 0.4 m.

The main part of the deck and the edge cantilevers are both taken to be isotropic, and consequently, the only properties associated with the elements (other than their material properties) are their depths. The elements in the main part of the deck are given a depth of 1.2 m and those in the edge cantilevers a depth of 0.4 m.

This model was analysed by the authors under the action of self-weight. A plane FE model (in accordance with the recommendations of Chapter 5) and a three-dimensional FE model using solid brick elements were also analysed. Figure 7.18a shows the calculated longitudinal stress

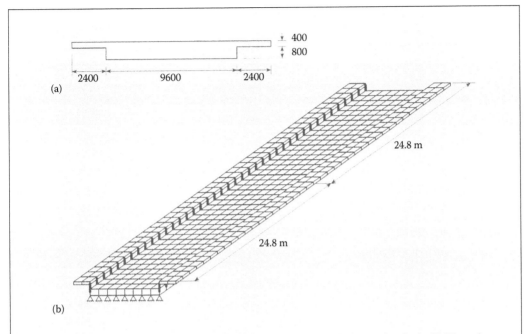

Figure 7.17 Upstand FE model of Example 7.4: (a) cross section (dimensions in mm); (b) FE mesh.

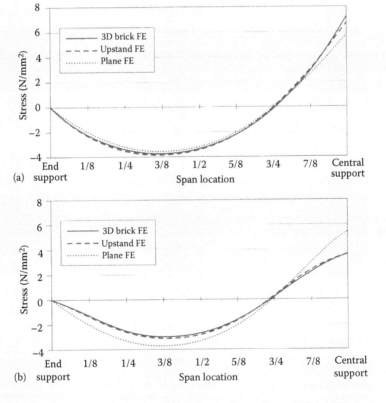

Figure 7.18 Longitudinal bending stress at top fibre for bridge of Example 7.4: (a) at centre; (b) 0.6 m in from edge of cantilever.

distributions at the top of the bridge deck along the centreline. As the model is symmetrical about the central support, only one span is shown in the figure. This stress distribution follows the expected pattern with zero stress at the ends, maximum compressive stress close to 3/8 span, zero stress close to 3/4 span and maximum tensile stress above the central support. The three-dimensional brick FE and upstand FE models predict very similar stress distributions at all locations, and the plane FE model is in reasonable agreement.

Figure 7.18b illustrates the corresponding distribution along a line 0.6 m in from the edge of the cantilever. The three-dimensional brick FE and upstand FE models once again predict very similar stress at all locations, but the plane FE model is in poor agreement with these. The plane FE model predicts a significantly greater stress at both the mid-span and central support locations. This is caused by the inability of the planar model to allow for the rising neutral axis in the edge cantilever. Alternatively, this can be viewed as the inability of the planar model to allow for shear lag. This example shows the benefits of three-dimensional modelling over planar modelling for bridge decks of this type.

When interpreting the results of an upstand FE model, it is important to realise that the moments are not comparable to those in a planar FE, as bending in the upstand model is not about the bridge neutral axis. However, the difference is accounted for by the presence of axial forces, which the bridge must be designed to resist.

7.5.1 Upstand finite element modelling of voided slab bridge decks

The three-dimensional nature of upstand FE modelling requires the specification of the correct area for the elements as well as the correct second moment of area, as the stiffness of each part of the deck is made up of a combination of both of these. Most FE programs only allow the specification of a depth for the FEs, which does not allow the independent specification of area and second moment of area. This is sufficient when dealing with solid slabs but causes problems when dealing with voided slabs.

Modelling of voided slabs by the planar FE method is discussed in Section 6.4. When considering the longitudinal direction, the depth of the FEs is determined by equating the second moment of area of the voided slab to that of an equivalent depth of solid slab. As the voids are generally located close to mid-depth of the slab, the equivalent depth of the elements will generally be quite close to (but smaller than) the actual depth of the voided slab. In other words, the presence of the voids does not greatly affect the longitudinal second moment of area of the deck. This is not the case when considering the cross-sectional area, which is greatly reduced by the presence of the voids. Therefore, an FE with a depth chosen by considering the second moment of area of the voided slab will have an excessive area.

As stiffness in the upstand FE model is made up of a combination of both the second moment of area and the cross-sectional area of the elements, this will result in an overly stiff model. A solution to this problem is to reduce the area of the elements. In theory, this could be done by incorporating additional beam members into the model with a negative area and zero second moment of area. Clearly, a member with negative area has no physical meaning, and quite sensibly, most computer programs will not allow this. A more feasible alternative is to choose the depth of the FEs so that they have the correct area and then to add additional beam members to make up the shortfall in the second moment of area. The additional beam members should have zero (or negligible) area. They should also have zero in-plane second moment of area as the in-plane behaviour is still modelled by the FEs.

EXAMPLE 7.5   UPSTAND FINITE ELEMENT MODEL OF VOIDED SLAB

Figure 7.19a shows the cross section of a voided slab bridge deck with wide edge cantilevers. The deck is simply supported with a 24 m span and is supported continuously across its breadth at each end. An upstand FE model is required.

Figure 7.19b shows the cross section of a suitable upstand FE model for this bridge deck. A choice of 20, 1.2 m long elements in the longitudinal direction would be appropriate for this model. The length of the rigid vertical members is equal to the distance between the centroid of the cantilevers and that of the main part of the deck. In this case, the vertical members are 0.35 m long.

The X direction is chosen as the longitudinal direction. For the elements in the main part of the deck, each element represents a portion of deck 1.2 m wide with one void. The second moment of area of this is

$$I_x = \frac{1.2(1.2)^3}{12} - \frac{\pi(0.7)^4}{64} = 0.161\,m^4$$

and the area is

$$A_s = (1.2)^2 - \frac{\pi(0.7)^2}{4} = 1.055\,m^2$$

Equating this to an equivalent solid element with the same area gives an equivalent element depth, d_{eq}, of

$$d_{eq} = \frac{1.055}{1.2} = 0.879\,m$$

The second moment of area of this equivalent solid element, I_{eq}, is

$$I_{eq} = \frac{1.2(0.879)^3}{12} = 0.068\,m^4$$

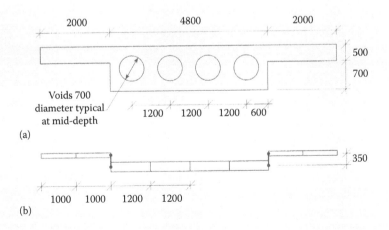

Figure 7.19 Upstand FE model of Example 7.5 (dimensions in mm): (a) cross section; (b) section through FE model.

This gives a shortfall in the second moment of area, which has to be made up by additional beam members. The second moment of area of these additional members, I_x^{add}, is

$$I_x^{add} = 0.161 - 0.068 = 0.093 \text{ m}^4$$

To incorporate the additional members in the model, each FE in the main part of the deck is replaced by four elements and four beam members, as illustrated in Figure 7.20. These elements have the same equivalent depth of 0.879 m, and the longitudinal beam members have second moments of area of 0.093 m⁴.

The required transverse second moment of area per unit breadth is given by Equation 6.1:

$$i_y^{v\text{-slab}} = \frac{d^3}{12}\left[1 - 0.95\left(\frac{d_v}{d}\right)^4\right]$$

$$= \frac{1.2^3}{12}\left[1 - 0.95\left(\frac{0.7}{1.2}\right)^4\right]$$

$$= 0.128 \text{ m}^4/\text{m}$$

Hence, the required additional second moment of area that is provided by the transverse beam members is

$$I_y^{add} = 1.2 \times 0.128 - 0.068 = 0.086 \text{ m}^4$$

The edge cantilevers are modelled as FEs with a depth of 0.5 m, which is equal to the actual depth of the cantilever. Rigid or very stiff vertical beam members are specified at 0.6 m intervals to join the meshes on the different planes. The elements used for this example only had nodes at the corners with the result that they could only be joined to the vertical members at their corners. Therefore, the originally proposed 1.0 × 1.2 elements in the cantilever were replaced with four 0.5 × 0.6 elements to give nodes at 0.6 m intervals. The final upstand FE model with beam members shown as dark lines is illustrated in Figure 7.21. Only one half of the model is shown as it is symmetrical.

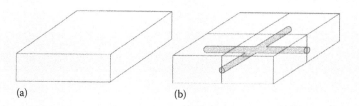

(a)   (b)

Figure 7.20 Replacement of plate element: (a) original element; (b) corresponding combination of plate and beam elements.

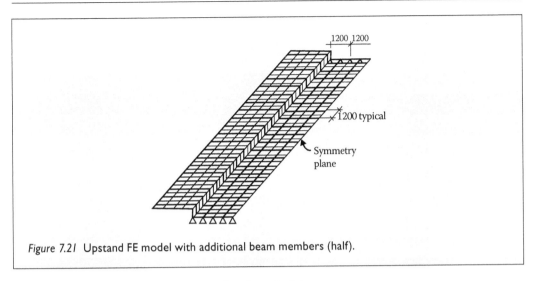

Figure 7.21 Upstand FE model with additional beam members (half).

7.5.2 Upstand FE modelling of other bridge types

It is possible to extend the principles of upstand FE analysis to types of bridges other than solid and voided slabs, provided care is taken to ensure that good similitude exists between the model and the actual structure. Figure 7.22a shows a beam-and-slab bridge. Each beam in this bridge will act compositely with the slab above it, and they are normally assumed to bend about their own centroid rather than that of the bridge as a whole. However, this is clearly an approximation as the exact location of the neutral axis will depend on the flange widths and the relative stiffnesses of the members. In such cases where the location of the neutral axis is unclear, an upstand FE analysis can be used to represent the behaviour more accurately than the alternative planar models.

The slab can be represented in the model using FEs located at its centroid and of equal depth to it. The properties of the remaining parts of the deck are then calculated, each about its own centroid, and are represented by beam members at the levels of those centroids, as illustrated in Figure 7.22b. The horizontal members at different levels are joined by stiff vertical members. This approach has the advantage of simplicity as there is a direct correspondence between each member and a part of the structure. However, the interpretation of the output can be tedious. The calculated moment for each beam member is only applicable to bending about its own centroid. If reinforcement is to resist the stresses in a beam and the adjacent slab, then the total moment will have to be calculated, taking account of the axial forces in the beam and the elements and the distance between them.

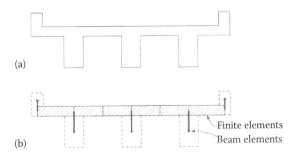

Figure 7.22 Upstand FE model of beam-and-slab bridge: (a) cross section; (b) section through upstand FE model.

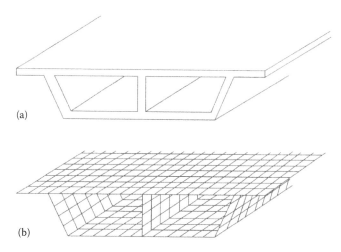

Figure 7.23 Plate FE model of cellular bridge: (a) original bridge; (b) FE model.

Figure 7.23a shows a cellular bridge deck and Figure 7.23b shows a suitable model based on a variation of the upstand FE analogy. This model, as well as dealing with a varying neutral axis, has the advantage of automatically allowing for transverse cell distortion as discussed in Section 6.6. Transverse diaphragms could also be incorporated into this model with ease. Care should be taken with such a model to ensure that sufficient numbers of elements are provided through the depth of the webs to correctly model longitudinal bending there.

7.5.3 Prestress loads in upstand FE models

When analysing for the effects of prestress in bridge decks, it is usual to uncouple the in-plane and out-of-plane behaviours. The in-plane behaviour is governed by the distribution of axial stress in the bridge deck and is often determined by a hand calculation. The out-of-plane behaviour is affected by the vertical components of tendon force and by the moments induced by tendon eccentricity. These effects are generally dealt with by calculating the equivalent loading due to prestress (Chapter 2), which is often based on an assumed location for the neutral axis.* The bridge deck is then analysed to determine the effects of the equivalent loading. The stresses determined from this analysis are combined with the in-plane axial stresses to obtain the overall effect of the applied prestressing forces.

When using an upstand FE method, the equivalent loading due to prestress can be applied in a three-dimensional manner. Many of the complications involved in determining equivalent loads due to prestress can be avoided in this way. There is no uncertainty concerning the location of the axis about which eccentricity of prestress must be calculated. There are also advantages to be gained in the interpretation of results, because they can be related directly to the design, without the need to distinguish between primary and secondary (parasitic) effects. This method is often simpler to implement as there is no need to uncouple

---

* Throughout this book, 'neutral axis' is used to mean the axis about which a segment of slab bends. For prestressed concrete, neutral axis is still taken to mean the axis about which the segment bends, even if this is not a point of zero stress in the presence of the axial prestress force.

the in-plane and out-of-plane behaviours. In the three-dimensional approach, the prestress forces are applied directly to the model at the correct vertical location by means of stiff vertical beam members. It follows that the calculation of moments due to cable eccentricity is not dependent on any assumed neutral axis location.

The sources of error in a traditional planar model, with the equivalent loading calculated in the normal manner, are two-fold. Firstly, as discussed in previous sections, the inability of the planar model to allow for the variation in neutral axis location may cause inaccuracies in the calculated response to equivalent loading. However, there is an additional error as the magnitude of the equivalent loading is itself dependent on the eccentricity of prestress and is therefore affected by the neutral axis location. It should be mentioned that, as the neutral axis location is load dependent, the location that is applicable to, say, self-weight may not be applicable to prestressing.

Figure 7.24a shows a portion of a bridge deck with an edge cantilever. The location of the neutral axis is indicated in Figure 7.24a, but it is unknown at this stage. The deck is subjected to a prestress force, P, at a distance, h, below mid-depth of the main part of the deck. This prestress force has an unknown eccentricity, e, which is also indicated in the figure. Figure 7.24b shows the equivalent portion of an upstand FE model. The prestress force is applied directly to the model through a rigid vertical member of length h. The eccentricity

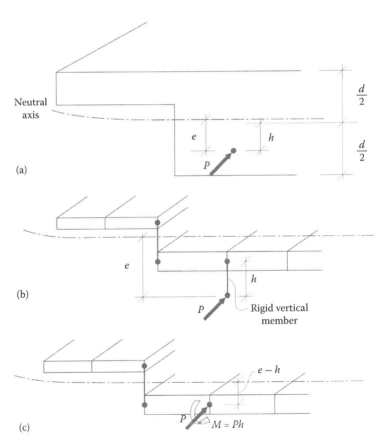

Figure 7.24 Portion of prestressed concrete deck: (a) original deck; (b) upstand FE model with vertical beam at point of application of prestress; (c) alternative upstand FE model.

of this force is once again *e*, but knowledge of the magnitude of the eccentricity is not necessary. The model is subjected to an axial force, which generates a moment of

Moment = Pe

To avoid the necessity of adding a large number of vertical beam members to the model, the prestress force can alternatively be applied at the level of the elements along with an additional moment to allow for the difference in level between the true point of application and the element. Figure 7.24c shows this alternative model. The additional moment is the product of the prestress force and the distance *h*. The equivalence of Figure 7.24a and c can be seen by considering the applied moment. In the latter, the applied moment is

Moment = $P(e - h) + Ph = Pe$

which is equal to the applied moment of the former. In this way, the independence of the prestress loading from the neutral axis location is retained, but the necessity of a large number of vertical members is avoided.

The authors have found this direct method of representing the effects of prestress to be the most accurate of many methods tested, when compared to results from three-dimensional brick FE analyses.

Chapter 8

Probabilistic assessment of bridge safety

8.1 INTRODUCTION

A common problem among bridge owners/managers is the need to reduce spending while attempting to operate and maintain an increasingly ageing bridge stock. In response to this challenge, the past decade has seen increased interest by bridge owners and managers in the use of probabilistic methods for the assessment and management of their bridges. Probabilistic assessments are usually employed once a deterministic assessment of a structure has deemed it necessary to repair, rehabilitate or replace it. They have been shown to provide significant cost savings by demonstrating that structures are safe, which would otherwise have been repaired or replaced.

The basic principle of a probabilistic approach is to derive a bridge-specific code and consequently a bridge-specific safety rating, hereafter termed the reliability index, β. This involves the statistical modelling of load and resistance parameters obtained by consulting the original design basis, from on-site measurements and/or from as-built drawings. Statistical updating of these models may be performed where additional information is available. The first stage of the process involves deterministic assessment to identify the critical limit state (LS) for the structure (see Figure 8.1). This deterministic assessment is performed according to generalised rules, in a traditional assessment process, using generalised partial safety factors provided by assessment codes. Where the results of this traditional assessment fail to demonstrate adequate safety/capacity, then the options available to the bridge owner/manager are (i) to implement a traditional strengthening project or (ii) to perform a bridge-specific probabilistic assessment. The latter approach, which is free from deterministic generalisations, computes a formal probability of failure, P_f, for the structure at the considered critical LS. The probability of failure, P_f, is related to the reliability index, β, by the following equation:

$$\beta = -\Phi^{-1}(P_f) \tag{8.1}$$

where Φ is the cumulative distribution function of the standardised normal distribution. Comparison of β with the target value specified in codes of practice, β_{target}, and demonstration that it exceeds these requirements ($\beta \geq \beta_{\text{target}}$) are deemed sufficient to validate the safety of the structure. Where, however, the result of the probabilistic assessment fails to demonstrate adequate safety ($\beta < \beta_{\text{target}}$), then the options open to the bridge owner/manager are to either (i) consider reducing the structure's classification (e.g. impose a load limit on the structure) or (ii) institute a strengthening plan developed using probabilistic models.

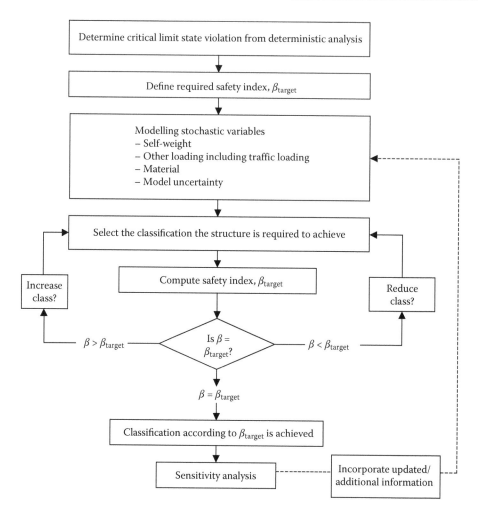

Figure 8.1 Probabilistic assessment process.

It is important to stress that at no stage is the safety of the structure compromised; rather, it is made more consistent across the stock of bridge structures through the provision of bridge-specific 'individualised' safety ratings. This is free from the imposed generalisations of deterministic assessment codes, which must necessarily be conservative to be widely applicable. Additionally, at any stage in the process, information from structural health monitoring (SHM) sensors or weigh-in-motion systems may be incorporated into the models.

8.2 CODE TREATMENT OF PROBABILITY OF FAILURE

Codes of practice define the probability of failure, P_f, in terms of a target reliability index, β_{target}, as detailed, for example, in Table 8.1.

In bridge design and assessment, the target reliability level, β_{target}, is the level of reliability prescribed by the bridge owner/manager to ensure acceptable safety and serviceability of the element/structure/network analysed. The choice of the target level of reliability should therefore take into account the possible consequences of failure in terms of risk to life or

Table 8.1 Relationship between β and P_f

P_f	10^{-1}	10^{-2}	10^{-3}	10^{-4}	10^{-5}	10^{-6}	10^{-7}
β_{target}	1.3	2.3	3.1	3.7	4.2	4.7	5.2

Source: SAMCO, *Final Report – F08a Guideline for the Assessment of Existing Structures*, Structural Assessment Monitoring and Control, 2006. Available from http://www.samco.org/.

injury, potential economic losses and the degree of societal inconvenience. It should also consider the expense and effort required to reduce the risk of failure. In the following subsections, the levels specified by a selection of different codes of practice and guidance documents are presented.

8.2.1 Eurocode 1990

The Eurocodes define consequence classes (CCs) that take account of the consequences of failure. These CCs are presented in EN 1990 (2002) and are given here in Table 8.2.

The three CCs can be associated with reliability classes (RCs). Table 8.3 gives the recommended target reliability levels at the ultimate LS (ULS).

The two β-values in Table 8.3 for each RC correspond approximately to the same reliability level. For example, if one considers a structure with RC3, then a reliability index $\beta_{target} = 4.3$ should be employed, provided that the probabilistic models of the basic variables are related to a reference period of 50 years. The same reliability level is reached where $\beta_{target} = 5.2$ is applied using models for 1 year. Considering a reference period equal to the remaining working life, the reliability level may be computed as (EN 1990 2002):

$$\beta_{t\_REF} = \Phi^{-1}\{[\Phi(\beta_1)^{t\_REF}]\} \tag{8.2}$$

Table 8.2 Definition of CC

Consequence class	Description	Examples of buildings and civil engineering works
CC3	*High* consequence for loss of human life, or economic, social or environmental consequences *very great*	Grandstands, public buildings where consequences of failure are high (e.g. a concert hall)
CC2	*Medium* consequence for loss of human life, economic, social or environmental consequences *considerable*	Residential and office buildings, public buildings where consequences of failure are medium (e.g. an office building)
CC1	*Low* consequences for loss of human life and economic, social or environmental consequences *small or negligible*	Agricultural buildings where people do not normally enter (e.g. storage buildings, greenhouses)

Source: EN 1990, *Eurocode – Basis of Structural Design*. European Committee for Standardisation, Brussels, 2002.

Table 8.3 EN 1990:2002 target reliabilities

Reliability class	Minimum values for β_{target}	
	1-year reference period	*50-year reference period*
RC3	5.2	4.3
RC2	4.7	3.8
RC1	4.2	3.3

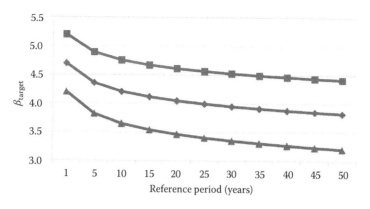

Figure 8.2 β_{target} versus Reference period.

where β_1 = target reliability index taken from Table 8.3 for a relevant RC and a reference period of 1 year (Sykora and Holicky 2011). Figure 8.2 illustrates computed values of β_{target} for varying reference periods.

8.2.2 ISO/CD 13822:2010

In ISO/CD 13822:2010, the target reliability is specified as a function of the type of LS examined as well as on the consequences of failure. As Table 8.4 shows, for the serviceability limit state (SLS), the target reliability index, β_{target}, varies from 0.0 for reversible states to 1.5 for irreversible states. At the ultimate limit state (ULS), β_{target} varies from 2.3 for very low consequences of a structural failure to 4.3 for structures whose failure would have very high consequences. β_{target} values for fatigue limit states (FLS) are specified according to the inspectability of the element. Interestingly, the reference period is also seen to vary according to the LS considered.

Table 8.4 ISO/CD 13822:1999 target reliabilities

Limit states	Target reliability index β_{target}	Reference period
Serviceability		
Reversible	0.0	Intended remaining working life
Irreversible	1.5	Intended remaining working life
Fatigue		
Inspectable	2.3	Intended remaining working life
Not inspectable	3.1	Intended remaining working life
Ultimate		
Very low consequences of failure	2.3	L_s years[a]
Low consequences of failure	3.1	L_s years[a]
Medium consequences of failure	3.8	L_s years[a]
High consequences of failure	4.3	L_s years[a]

[a] L_s is a minimum standard period of safety (e.g. 50 years).

Table 8.5 NKB target reliabilities for ULS

Failure type	Failure type I, ductile failure with remaining capacity	Failure type II, ductile failure without remaining capacity	Failure type III, brittle failure
Failure consequences			
Less serious	3.1	3.7	4.2
Serious	3.7	4.2	4.7
Very serious	4.2	4.7	5.2

8.2.3 Nordic Committee on Building Regulations

The Nordic Committee on Building Regulations (NKB) Report No. 36 (NKB 1978) gives reliability indices, again as a function of the failure type and consequence. The type of failure is assessed on the basis of the characteristics for the given material, component or structure, that is, ductile or non-ductile. The values recommended for the ULS for a reference period of 1 year are given in Table 8.5. For the SLS, NKB recommends values of $\beta = 1.0$ to 2.0. It is noted that the values presented in Table 8.5 form the basis for the World Road Association report on reliability assessment of bridges (PIARC 1999).

8.2.4 International Federation for Structural Concrete Bulletin 65

The International Federation for Structural Concrete (fib) Bulletin 65 Model Code 2010 (fib 2012) provides the recommendations of Table 8.6 for target reliabilities for the design of structures.

The fib Bulletin 65 suggests that the β_{target} values given in Table 8.6 may also be used for the assessment of existing structures but suggests that a differentiation of the target reliability level for new structures and for existing structures may need to be considered. However, the decision to choose a different target reliability level for existing structures may be taken only on the basis of well-founded analysis of the consequences of failure and that the cost of safety measures for any specific case may need to be considered. Suggestions for the reliability indices for existing structures are given in Table 8.7 for the specified reference periods.

Table 8.6 fib Bulletin 65 target reliabilities

Limit states	Target reliability index, β_{target}	Reference period
Serviceability		
Reversible	0.0	Service life
Irreversible	1.5	50 years
Irreversible	3.0	1 year
Ultimate		
Low consequence of failure	3.1	50 years
	4.1	1 year
Medium consequence of failure	3.8	50 years
	4.7	1 year
High consequence of failure	4.3	50 years
	5.1	1 year

Table 8.7 Target reliabilities for existing structures

Limit states	Target reliability index β	Reference period
Serviceability	1.5	Residual service life
Ultimate	In the range of 3.1–3.8[a]	50 years
	In the range of 3.4–4.1[a]	15 years
	In the range of 4.1–4.7[a]	1 year

[a] Depending on costs of safety measures for upgrading the existing structure.

8.2.5 AASHTO

According to the *AASHTO LRFD Design Specifications* (AASHTO 2012), the Strength I LS has been calibrated for a target reliability index of 3.5 with a corresponding probability of exceedance of 2.0×10^{-4} for a 75-year design life of the bridge. This is equivalent to an annual probability of exceedance of 2.7×10^{-6} with a corresponding annual target reliability index of 4.6. At the time of writing, calibration of target reliabilities for the service LSs is awaiting completion.

8.3 CALCULATION OF THE PROBABILITY OF FAILURE, P_f

The reliability of an engineering system refers to the probability of survival or its complement, the probability of failure (Melchers 1999). The concept of a 'LS' is used to define 'failure' in the context of structural reliability analysis. The term 'failure' does not necessarily imply structural collapse but, in most cases, refers to a situation when the performance of the structure exceeds a predefined limit. For example, if the LS to be considered is the initiation of reinforcement corrosion, then failure in this case may be defined as when the chloride content at the reinforcement depth exceeds a critical value, C_{cr}. Thus, the LS is a boundary between desired and undesired performance of a structure and defines a failure surface as indicated in Figure 8.3.

In a standard case, when the LS function is not time-dependent, it may be written as

$$Z = G(R, S) = R - S \tag{8.3}$$

where Z is the LS margin (e.g. the excess of resistance over load effect), G is the LS function, R is the random variable representing the resistance and S is the random variable representing the corresponding load effect or action. R and S may be functions of other variables,

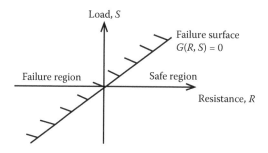

Figure 8.3 LS Schematic. (With kind permission from Springer Science+Business Media: *Structural Reliability Theory and its Applications*, 1982, Toft-Christensen, P. and Baker, M.J.)

deterministic or random. Both R and S may also be functions of time, t, as schematically shown in Figure 8.4 (Stewart and Rosowsky 1998).

If in the LS function, $Z < 0$, then the failure state is reached. The probability that the LS has been violated ($Z < 0$) is referred to as the probability of failure (P_f) and can be obtained by integrating the probability density function (PDF) (Melchers 1999):

$$P_f = P[G(\boldsymbol{X}) \leq 0] = \int \cdots \int_{G(\boldsymbol{X}) \leq 0} f_X(x)\, \mathrm{d}x \tag{8.4}$$

where $\boldsymbol{X}$ is the vector of all relevant basic variables, $G(\boldsymbol{X})$ is the LS function that expresses the relationship between the LS and the basic variables and $f_X(x)$ is the joint PDF for the n-dimensional vector $\boldsymbol{X}$ of basic variables.

The region of integration $G(\boldsymbol{X}) \leq 0$ denotes the space of LS violation. Theoretically, the solution to Equation 8.4 can be obtained through one of the following three methods (Melchers 1999):

1. Direct analytical integration
2. Numerical integration such as simulation methods
3. By transforming the integrand into a multi-normal joint PDF for which some solutions exist

Except for some special cases, the integration of Equation 8.4 over the failure domain $G(\boldsymbol{X}) \leq 0$ cannot be performed analytically. In addition, the LS equation contains functions of the basic variables, which are too complicated for calculus to be used in the evaluation of their integrals. In such cases, methods 2 and 3 become practical choices for the evaluation of the failure probability. One common technique of type 2 methods is Monte Carlo simulation. As for type 3 methods, there are several classical techniques such as the first-order reliability method (FORM) and the second-order reliability method (SORM). In the FORM approach, the failure surface is approximated as linear as discussed in Section 8.3.1. However, there are cases where the LS function is more curved, and as such, approximation by a linear function becomes less accurate. In such cases, the curvature itself is expected to have an influence on the definition of any simple surface employed to approximate the non-linear LS function (Melchers 1999). In this situation, the SORM approach, developed to deal with the non-linearity of the LS function, is more appropriate. All of these methods

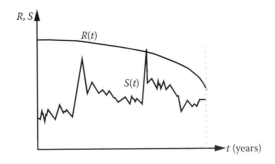

Figure 8.4 Schematic representation of time-variant reliability problem (R = resistance, S = load effect, t = time). (From Stewart, M.G. and Rosowsky, D.V., *Journal of Infrastructure Systems*, 4(4), 146–155, 1998.)

are described in detail in the literature. For a more in-depth treatment of their theoretical basis, the reader is referred to Melchers (1999), Haldar and Mahadevan (2000), Madsen et al. (1986), Ang and Tang (1984), Thoft-Christensen and Baker (1982) and Schneider (1997).

8.3.1 Basic statistical concepts

Some basic statistical concepts are commonly used in reliability analysis. If a random variable, Z, is a linear combination of n independent, random variables, $X_1, X_2,..., X_n$:

$$Z = a_0 + a_1 X_1 + a_2 X_2 + \cdots + a_n X_n \tag{8.5}$$

then the mean is given by the corresponding linear combination of means:

$$\mu_Z = a_0 + \sum_{i=1}^{n} a_i \mu_{X_i} \tag{8.6}$$

If the random variables are independent, the standard deviation of Z is given by

$$\sigma_Z = \sqrt{a_1^2 \sigma_{X_1}^2 + a_2^2 \sigma_{X_2}^2 + \cdots + a_n^2 \sigma_{X_n}^2} \tag{8.7}$$

If, on the other hand, they are correlated, the standard deviation is given by

$$\sigma_Z = \sqrt{a_1^2 \sigma_{X_1}^2 + a_2^2 \sigma_{X_2}^2 + \cdots + a_n^2 \sigma_{X_n}^2 + \sum_{i=1}^{n} \sum_{j=1, j \neq i}^{n} \rho_{ij} a_i a_j \sigma_{X_i} \sigma_{X_j}} \tag{8.8}$$

where ρ_{ij} is the correlation coefficient between X_i and X_j. The case where both the load effect S and resistance R are independent and normally distributed as $N(\mu_S, \sigma_S)$ and $N(\mu_R, \sigma_R)$, respectively, is illustrated in Figure 8.5a.

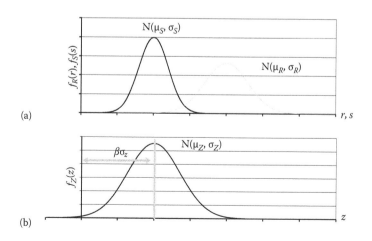

Figure 8.5 Reliability index. (a) Normal distributions of load (S) and resistance (R). (b) Normal distribution of LS margin (Z).

The LS margin, Z, is described by Equation 8.3 where

$$\mu_Z = \mu_R - \mu_S \qquad (8.9)$$

and from Equation 8.7,

$$\sigma_Z = \sqrt{\sigma_R^2 + \sigma_S^2} \qquad (8.10)$$

Since R and S are normal, Z, a linear function of R and S, is also normally distributed, $N(\mu_Z, \sigma_Z)$ as illustrated in Figure 8.5b. The Cornell reliability index, β, is defined as the distance from the mean point to the LS surface (PIARC 1999). This distance, measured in units of standard deviation, is defined in Equation 8.11 and is illustrated in Figure 8.5b wherein positive values of Z represent a safe domain, whereas negative values represent the unsafe domain, and zero values indicate the LS boundary.

$$\beta = \frac{\mu_Z}{\sigma_Z} \qquad (8.11)$$

This definition may be considered as exact for linear safety margins of normally distributed variables. However, for non-linear safety margins, it is necessary to employ approximation methods such as Taylor's expansion series (Melchers 1999) as previously discussed. For the general case, a lack of invariance, whereby equivalent margins can give different results depending on the way in which they are formulated, is a problem for Equation 8.11; a different answer can be computed depending upon whether the applied moment is compared to the bending capacity or the applied load is compared to the plastic collapse load (PIARC 1999). In these cases, the general solution provided by Hasofer and Lind (1974) provides a methodology to evaluate the reliability index. In this case, the basic random variables (X) are transformed into independent standard normal variables (U) through approximate transformations (Figure 8.6).

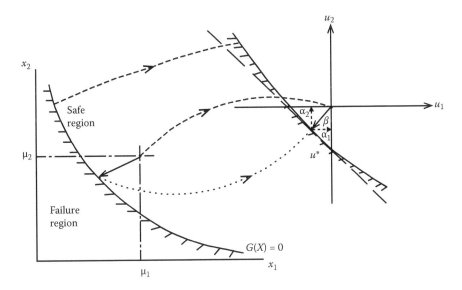

Figure 8.6 Transformation to the standard normal space. (From PIARC, Reliability-Based Assessment of Highway Bridges, Technical Committee 11 Bridges and Other Structures, World Road Association, 1999.)

Where the normal random variables are uncorrelated, this transformation can be performed without complication. However, if this is not the case, the random variable set, which is uncorrelated, must be determined, and this new set can then be transformed (Melchers 1999). The properties of the U-space are such that the geometrical distance from the origin to any point on the transformed failure surface corresponds to the distance between the mean value point and the associated point on the original failure surface (Figure 8.6). The reliability index in the special U-space is therefore defined as the shortest distance from the origin to the failure surface and is referred to as the Hasofer–Lind reliability index (PIARC 1999). The point in both the U-space and the X-space, which corresponds to the Hasofer–Lind reliability index, is termed the *design point*, u^*. This is the point of greatest probability density or the point of 'maximum likelihood' for the failure domain (Melchers 1999). The coordinates of the design point are

$$u_i^* = -\alpha_i \beta \tag{8.12}$$

where α_i, illustrated in Figure 8.6, are the direction cosines, which describe the weighted contributions of the variables to β (Thoft-Christensen and Baker 1982; Madsen et al. 1986; Melchers 1999).

EXAMPLE 8.1   COMPUTATION OF β FOR INDEPENDENT AND NORMALLY DISTRIBUTED VARIABLES

Consider the case where both the load effect S and resistance R are independently and normally distributed as $N(10, 2)$ and $N(20, 2)$, respectively. Find the reliability index β.

The LS margin, Z, therefore has a mean value $\mu_z = 20 - 10 = 10$ kN and standard deviation $\sigma_z = \sqrt{\sigma_R^2 + \sigma_S^2} = 3.61$ kN. The reliability index is therefore calculated as $\beta = \mu_z / \sigma_z = 2.77$.

EXAMPLE 8.2   COMPUTATION OF β FOR MORE THAN TWO INDEPENDENT AND NORMALLY DISTRIBUTED VARIABLES WITH LINEAR LS APPROXIMATION

For the statically determinate structure illustrated in Figure 8.7, determine (a) the value of the reliability index β, (b) the weighting factors, α_i, associated with each variable at the design point

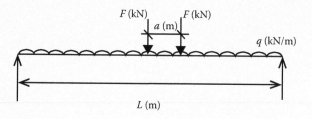

Figure 8.7 Statically determinate system.

Table 8.8 Parameters of variables

	μ	σ
R	4000	400
F	200	20
q	27	1

(β-point) and (c) the values of the governing variables at the β-point. The load (F and q) and resistance (R) variables are normally distributed with the parameters given in Table 8.8. The span length, L, and axle spacing are considered as deterministic with $L = 15$ m and $a = 1.2$ m. The variables can be assumed to be uncorrelated.

a. The LS margin is evaluated at the mid-point of the span, that is, the location of maximum bending moment:

$$Z = R - S$$

$$= R - F\left(\frac{L}{2} - \frac{a}{2}\right) - \frac{qL^2}{8}$$

$$= R - 6.9F - 28.125q$$

Hence, the LS is a linear function of the variables R, F and q. The mean is therefore

$$\mu_z = \mu_R - 6.9\mu_F - 28.125\mu_q$$

$$= 4000 - 6.9(200) - 28.125(27)$$

$$= 1861\,\text{kNm}$$

and the standard deviation is

$$\sigma_z = \sqrt{\sigma_R^2 + 6.9^2\sigma_F^2 + 28.125^2\sigma_q^2} = \sqrt{(400)^2 + 47.61(20)^2 + 791.02(1)^2} = 424\,\text{kNm}$$

Hence, $\beta = \mu_z/\sigma_z = 4.39$.

b. The weighting factors, α_i, describe the relative contributions of each variable to the β value at the design point. They are calculated as (Schneider 1997)

$$\alpha_i = a_i \frac{\sigma_i}{\sigma_z} \tag{8.13}$$

where a_i are the coefficients of the random variables and

$$\sum_{i=1}^{n} \alpha_i^2 = 1.0 \tag{8.14}$$

Thus,

$$\alpha_R = (1)\frac{400}{424} = +0.943$$

$$\alpha_F = (-6.9)\frac{20}{424} = -0.325$$

$$\alpha_q = (-28.125)\frac{1.0}{424} = -0.066$$

This shows that the largest component of the variability comes from the resistance, a significant amount comes from F and the contribution of q is relatively small. It is worth noting that a positive value for α_i implies that the variable increases the reliability index β, whereas a negative α_i implies that the variable reduces β. Hence, generally, resistance variables increase β, whereas load variables reduce β. An exception to this rule occurs for example in the case of arch bridges where loading can also act to increase structural capacity.

c. The value of a variable x_i^* at the β-point may be estimated using

$$x_i^* = \mu_i - \alpha_i \beta \sigma_i \qquad (8.15)$$

Hence,

$$x_R^* = 4000 - (0.943)(4.39)(400) = 2344.1 \, kNm$$

$$x_F^* = 200 - (-0.325)(4.39)(20) = 228.6 \, kNm$$

$$x_q^* = 27 - (-0.066)(4.39)(1) = 27.3 \, kNm$$

8.4 RESISTANCE MODELLING

When performing a safety assessment, it is important to accurately model the resistance variable, R, both at a point in time and as a function of time (Figure 8.4). This process requires information on the material properties, such as strength and stiffness, as well as the structure's dimensions. Sources of this information can include (i) the original codes/guidance documents to which the bridge was designed, (ii) steel mill certificates or concrete cylinder/cube tests from the time of construction or (iii) results from in situ destructive and/or non-destructive tests performed on the structure or from installed sensors. Model uncertainty should be included, and where tests have been carried out on the structure, this model uncertainty can be reduced via statistical updating techniques (NKB 1978; DRD 2004).

8.4.1 Reinforced concrete

Concrete

Material models for concrete must include the concrete compressive strength, f_c, the modulus of elasticity, E_c, the compressive strain and information on shrinkage and creep. COST345 (2004) identifies that the main sources of uncertainty in these concrete properties arise due to

- Variations in the properties of the concrete and proportion of concrete mix
- Variations in mixing, transporting, placing and curing methods
- Variations in testing procedures
- Concrete being in a structure rather than in test specimens

Concrete compressive strength is a very important parameter as it is generally included in models defining the load-carrying capacity of a concrete structure and is also often used as the basis variable for determining a number of other parameters (DRD 2004). Normal and lognormal distributions have both been used in the literature to represent the PDF of this parameter, although lognormal is generally preferred (PIARC 1999; COST345 2004).

When defining the mean value of the compressive strength for an existing structure, μ_{f_c}, it is vital that original documentation, including design codes from the time of the original design, are consulted. It is very important to know the relationship between the characteristic and mean strengths, and this relationship can be code-dependent (O'Connor and Enevoldsen 2008). The coefficient of variation (CoV) is generally higher for lower strength concrete. In DRD (2004), the CoV ranges from 12.0% for the higher strength concretes, that is, mean strengths from 47 to 58.7 N/mm² (characteristic values, f_{ck}, from 40 to 50 N/mm²), to 22.0% for mean strengths of the order of 6.76 N/mm² (f_{ck} = 5 N/mm²; Table 8.9). The values in Table 8.9 were derived for the characteristic value representing the 10% fractile on the basis of lognormal distribution of compressive strength.

The other material properties of concrete (tensile strength, modulus of elasticity and ultimate strain) can be determined from the compressive strength. The shrinkage and creep

Table 8.9 Distribution parameters for concrete compressive strength

f_{ck} (N/mm²)	μ_{f_c} (N/mm²)	CoV_{f_c}
5	6.76	0.22
10	12.8	0.18
15	18.9	0.17
20	24.8	0.16
25	30.6	0.15
30	36.2	0.14
35	41.7	0.13
40	47.0	0.12
45	52.8	0.12
50	58.7	0.12

Source: DRD, Report 291, Reliability-Based Classification of the Load Carrying Capacity of Existing Bridges, Road Directorate, Ministry of Transport, Denmark, 2004.

of the concrete can be determined by considering the available information on the age and geometry of the structure, the water/cement ratio of the concrete and the surrounding site climate. When assessing an existing structure, the age will usually be such that shrinkage and creep can be considered as having terminated. If they are to be included, the mean values of both shrinkage and creep can be determined using the approach in the fib Model Code 2010 (Fib 2012). Adopting that approach, the shrinkage strain can be taken as normally distributed with a CoV of 35.0%, and creep strain can also be taken as normally distributed with a CoV of 20.0% (DRD 2004).

The mean concrete strength can also be taken to have increased as a function of time. DRD (2004) allows for a 50% increase in compressive strength, from the 28-day value, for bridges built pre-1945. For bridges built after 1945, a 25% increase is assumed once the bridge is more than 5 years old. However, for bridges containing silica fume or accelerators, no increase is permitted. These values exclude any allowance for deterioration of the concrete, which may have reduced the concrete strength, or increased its CoV, as a function of time. In all cases, where possible, in situ strength testing is advised both to inform and update statistical distributions of governing variables.

Reinforcing steel

Probabilistic models of reinforcing steel strength are required to consider the variation in the strength of material, variation in the cross section, effect of rate of loading, effect of bar diameter on properties of the bar and effect of strain at which yield is defined (Mirza and MacGregor 1979; Wiśniewski et al. 2012). It is also shown that different tests can sometimes be performed to measure the same property. For example, yield strength recorded by the manufacturer in mill tests is approximately 8% greater than the actual static yield strength, so there are often two quoted steel strengths: the mill strength and the static strength. The mill strength tests are performed at a rapid rate of loading and use actual areas, whereas the static strengths are determined based on nominal area and use a rate of strain similar to that expected in a structure.

DRD (2004) suggests that the tensile yield strength, f_y, can be assumed to be lognormally distributed with a constant standard deviation of 25 N/mm^2, independent of the grade (see Table 8.10). The characteristic strength, f_{yk}, listed in Table 8.10 is the guaranteed yield strength.

Table 8.10 Tensile yield strength of non-prestressed reinforcement

Type	Symbol	Diameter (mm)	Characteristic strength, f_{yk}	Mean value, μ_{f_y} (N/mm^2)	Standard deviation, σ_{f_y}	CoV$_{f_y}$
Smooth bars	Fe 360	≤16	235	304	25	0.082
	Fe 360	≥16	225	293	25	0.085
	Fe 430	≤16	275	345	25	0.072
	Fe 430	≥16	265	334	25	0.075
	Fe 510	≤16	355	426	25	0.059
	Fe 510	≥16	345	416	25	0.060
Ribbed bars	Ks 410	–	410	482	25	0.052
	Ks 550	–	550	623	25	0.040
Cold deformed bars	T	–	550	623	25	0.040

Source: DRD, Report 291, Reliability-Based Classification of the Load Carrying Capacity of Existing Bridges, Road Directorate, Ministry of Transport, Denmark, 2004.

PIARC (1999) also suggests a lognormal distribution for the yield strength of steel. The lognormal distribution allows for the positive skewness of measured data and precludes negative values of strength. For high-strength steel, the suggested standard deviation is 30–35 N/mm², which corresponds with JCSS (2000b) which suggests 30 N/mm².

In terms of the cross section, the actual areas of the reinforcing bars can differ from the nominal areas due to the rolling process. COST345 (2004) suggests a normal distribution to represent this uncertainty. For groups of bars, PIARC (1999) suggests a lognormal random variable and presents an approach for calculating the resistance provided by a group of bars as a sum of the resistances of individual bars. In this case, the mean value and standard deviation (of the group of bars) are obtained as a function of individual bar characteristics.

PIARC (1999) also suggests that the effective depth (distance from the compressive face of the section to the centre of reinforcement) of the reinforcement is modelled as a random variable. This parameter can be affected by inaccuracies in slab thickness, height and spacing of supporting formwork or the diameters of the bars. While the mean values for the probabilistic distribution for this random variable can be taken as equal to the nominal value, the CoV will vary depending on the placement (i.e. top or bottom) and on possible deterioration. It can be in the range of 5%–20%. It is also suggested that the depth of cover to reinforcement should be taken as a random variable with a lognormal distribution (PIARC 1999).

The modulus of elasticity and the ultimate strain of the reinforcement are, however, often modelled deterministically. Such an assumption will not significantly affect the safety calculation (DRD 2004). The compressive strength of reinforcement can be determined from the tensile yield strength if no other information is available. If the reinforcement is not cold-formed, then they can be assumed to be equal, and, in the case of cold-formed reinforcement, it is suggested that the compressive yield strength may reasonably be taken as 0.8 times the corresponding tensile value (DRD 2004).

8.4.2 Prestressed concrete

In general, concrete standards do not give characteristic values for all material parameters of prestressed reinforcement. These values must therefore be based on documentation from the design or the manufacturer's documentation (DRD 2004).

In a probabilistic safety assessment involving a prestressed concrete bridge, the strength of the prestressing steel can be modelled as lognormally distributed variable (O'Connor and Enevoldsen 2008). In cases where the prestressing consists of wires, the system effect is taken into account in determining the strength parameters of the reinforcement. A low CoV, about 4.0%, is generally sufficient for prestressing steel (DRD 2004). The ultimate strain and the modulus of elasticity for the prestressing steel can be modelled deterministically without significantly affecting the safety calculations. The prestressing force at any given time should be determined, taking the relevant losses into account.

8.4.3 Structural steel

For the probabilistic model for the yield strength, f_y, of structural steel, a lognormal distribution is often recommended (JCSS 2000b; DRD 2004). The mean value is dependent on the steel grade and the thickness, t, and is greater than the characteristic value, f_{yk}. JCSS (2000b) proposes a probabilistic model, and DRD (2004) presents the recommended mean values for various grades of steel reproduced here in Table 8.11.

Studies differ on whether the standard deviation or the CoV of the yield strength should remain constant. DRD (2004) suggests a standard deviation of 25 N/mm² for all grades

Table 8.11 Yield strength of structural steel

Steel type	Characteristic value f_{yk} (N/mm²)			Mean value (N/mm²)			Standard deviation (N/mm²)
	$t \le 16$	$16 < t \le 40$	$40 < t$	$t \le 16$	$16 < t \le 40$	$40 < t$	All t
St 37	235	225	215	304	293	283	25
Fe 360	235	225	215	304	293	283	25
St 42 A	260	250	240	328	319	308	25
St 42, -1,-2,-3	260	250	240	328	319	308	25
St 44	275	265	255	344	334	324	25
St 42, B, C, D	275	265	255	344	334	324	25
Fe 430	275	265	255	344	334	324	25
St 50	340	330	320	410	400	390	25
St 52	340	330	320	410	400	390	25
Fe 510	355	345	335	426	416	405	25

Source: DRD, Report 291, Reliability-Based Classification of the Load Carrying Capacity of Existing Bridges, Road Directorate, Ministry of Transport, Denmark, 2004.

of steel, whereas JCSS (2000b) suggests a CoV of 7.0%. A lognormal distribution is also recommended for the ultimate tensile strength of structural steel. DRD (2004) recommends a constant standard deviation of 25 N/mm² for all steel grades, whereas JCSS (2000b) recommends a constant CoV of 4.0%. The modulus of elasticity, shear modulus and Poisson's ratio can be taken as deterministic or a lognormal distribution can be used with a small CoV (typically, 3.0% is suggested).

8.4.4 Soils

There is great uncertainty associated with soil parameters, and the geology of different sites can vary greatly, even within short distances. It is therefore generally necessary to base an evaluation of the relevant strength parameters for soil on geotechnical investigations and tests in the specific locality of the bridge. The uncertainty in the parameters can be determined on the basis of the guidelines in JCSS (2006).

8.4.5 Material model uncertainty

The model uncertainty takes account of (i) the accuracy of the calculation model, I_1, (ii) possible deviations from the strength of material properties in the structure involved as compared with that derived from control specimens, I_2, and (iii) the degree of control on site (i.e. materials identity), I_3 (NKB 1978). The model uncertainty is taken into account by introducing judgement factors, I_m, related to the material properties. The judgement factor, I_m, is assumed to be lognormally distributed with a mean value equal to 1.0 and CoV, V_{I_m}, which is introduced by multiplying the basic material variables by I_m. The CoV, V_{I_m} is calculated as (NKB 1978)

$$V_{I_m} = \sqrt{V_{I_1}^2 + V_{I_2}^2 + V_{I_3}^2 + 2\left(\rho_1 V_{I_1} + \rho_2 V_{I_2} + \rho_3 V_{I_3}\right) \cdot V_M} \qquad (8.16)$$

where the CoV V_{I_i} and the coefficient ρ_i, respectively, relate to (i) the accuracy of the calculation model, (ii) material property deviations and (iii) material identity and are as specified in Table 8.12, and V_M is the CoV of the basic material variable.

Table 8.12 Model uncertainty factors

Accuracy of calculation model			
	Good	*Normal*	*Poor*
V_{I_1}	0.04	0.06	0.09
ρ_1	−0.3	0.0	0.3

Material property deviations			
	Small	*Medium*	*Large*
V_{I_2}	0.04	0.06	0.09
ρ_2	−0.3	0.0	0.3

Material identity			
	Good	*Normal*	*Poor*
V_{I_3}	0.04	0.06	0.09
ρ_3	−0.3	0.0	0.3

Source: NKB, Report No. 36 Guidelines for Loading and Safety Regulations for Structural Design. Nordisk Komité for Bygningsbestemmelser, 1978.

Table 8.12 gives uncertainty factors associated with the accuracy of the calculation model, defining 'Good', 'Normal' and 'Poor' calculation accuracies. It should be noted that (DRD 2004; O'Connor and Enevoldsen 2008)

1. A Good computation model can for example be assumed where
   a. The model is so simple (corresponding to a simple structure) that only small variations can arise.
   b. Attention has been paid to eccentricities, secondary moments and so on.
   c. The model has been verified for the bridge in question.
   d. An improved model has resulted in a reduction of the uncertainty of an important stochastic variable.
2. A Normal calculation accuracy can be assumed in situations where the model is generally accepted as being in conformity with normal practice.
3. A Poor computation model is one that has been excessively simplified and does not meet the requirements for a model of normal accuracy.

The uncertainty associated with determining material parameters is dependent upon the amount of information available on the materials of the structure and on the availability of test results and so on. For uncertainty factors associated with material identity, it should be noted that (DRD 2004; O'Connor and Enevoldsen 2008)

1. Good material identity can be assumed if the materials have been verified for the bridge or if the identity of materials used subsequently can be documented (e.g. 'as built' drawings).
2. Normal material identity is assumed when the materials are assigned on the basis of the project material, and there is no reason to doubt that the bridge in question was not built in accordance with the project material.
3. Poor material identity arises when estimated values are used, or where the project material is dubious or information is incomplete.

EXAMPLE 8.3   COMPUTATION OF PARAMETERS INCLUDING MODEL UNCERTAINTY

For a given bridge, the concrete compressive strength is modelled as a lognormally distributed variable. It is specified in the original 1939 construction drawings with a 28-day characteristic cube strength = 20 N/mm². Determine the statistical parameters (i.e. μ, σ) of the material's distribution, including model uncertainty.

- Cylinder strength is 80% of cube strength: $f_{ck} = 0.8 \, f_{cu} = 16$ N/mm².
- For intact concrete structures, a deterministic increase in the compressive strength of the concrete with time can be assumed in the absence of contra-indications. The characteristic strength can be taken as 50% higher than the original 28-day strength if the bridge in question was constructed pre-1945 (DRD 2004).
- Thus, f_{ck} at 28 days may be increased by 50%. $1.5 f_{ck} = 24.0$ N/mm². From Table 8.9, $\mu_{f_c} = 29.4$ N/mm² and CoV $V_{f_c} \, (= V_M) = 15.2\%$.
- The accuracy of the calculation model is taken as Normal, the uncertainty of the material properties in construction is classified as Medium, whereas the material identity is classified as Normal. Thus, from Equation 8.16, $V_{I_m} = 10.39\%$.

Thus, the CoV including model uncertainty $V = \sqrt{V_M^2 + V_{I_M}^2} = 18.40\%$. The resulting standard deviation σ_{f_c} is (29.4 × 18.4/100 =) 5.41 N/mm².

8.5 DETERIORATION MODELLING

The prediction of bridge deterioration with time is a difficult task due to vast uncertainties in structural and environmental parameters (JCSS 2000a). Chloride-induced steel corrosion is one of the most widespread deterioration mechanisms for reinforced concrete (RC) structures worldwide. Tilly (2007) claims that it is in fact responsible for more than 55% of repairs carried out on structures across Europe, and, for that reason, it is the deterioration mechanism considered here (note that loss of bond and embrittlement are not considered). The main indicator of the presence of corrosion is cracking of the concrete cover. Generally, excessive cracking occurs before the corrosion has any significant effect on the overall structural capacity of the member. However, once cracks have developed, they can lead to rapid structural degeneration, and consequently, much work has been undertaken in recent years in an effort to accurately predict the time to initiation of these cracks.

The mechanism of chloride-induced corrosion is generally accepted to be electrochemical in nature. The alkaline environment of concrete results in the formation of a passive film at the steel–concrete interface, which protects the steel from corrosion. Chloride ions can destroy this passive film, and once the layer breaks down, corrosion is initiated. Due to the presence of both moisture and oxygen, a range of corrosive products (i.e. rust) are then formed. The corrosive products are greater in volume than the original steel; thus, the process is expansive and gives rise to tensile stresses in the concrete surrounding the reinforcing bar. The result is an acceleration in the rate of cracking and spalling of the cover layer.

The rate of ingress of chloride ions can be represented using Fick's second law:

$$C(x, t) = C_i + (C_s - C_i)\left[1 - \text{erf}\left(\frac{x}{2\sqrt{D_{\text{app}}t}}\right)\right]$$

(8.17)

where $C(x, t)$ is the total chloride content (percentage of Cl, by mass, of cement or concrete) at time t (years) of exposure and at depth x (mm) from the surface of the concrete; D_{app} is the apparent diffusion coefficient (mm²/year); C_s is the surface chloride content; erf is the Gaussian error function; and C_i is the initial chloride content often neglected. Figure 8.8 shows a typical chloride profile fitted to Fick's second law of diffusion (Kenshel 2009). In this figure, it can be seen that chloride content decreases with increasing depth from the surface of the concrete, except for the outer region. This is because the concrete skin has a different matrix composition compared to the internal concrete due to phenomena such as contact with the mould, segregation of aggregates or dielectric reaction between the concrete surface and the chloride environment (Andrade et al. 1997). Moreover, chlorides at the outer layer of the concrete cover can often be washed out by the rain or by the cooling water used during the sample extraction operation.

A conceptual model for service life prediction of corroded RC structures was first developed by Tutti (1982). As shown in Figure 8.9a, it consists of both an initiation phase, T_i, and a propagation phase, T_{cr} (Kenshel 2009). The first phase, T_i, represents the time required for the chloride ions to diffuse to the steel-concrete interface and the surface chloride concentration to exceed a limiting value, thereby initiating corrosion. The second phase, the propagation period, represents the time after corrosion initiation, as crack generation occurs. Further research carried out by Weyers et al. (1991) suggests that, in fact, not all of the corrosion products contribute immediately to the expansive pressure on the concrete. Instead, some are considered to fill the voids and pores around the reinforcing bar. This concept of a porous zone has resulted in the subdivision of the propagation period into three distinct and separate phases as shown in Figure 8.9b. The first, commonly referred to as T_{1st}, is the time to the appearance of first crack – generally taken to be a hairline crack of approximately 0.05 mm and represents the period of free expansion until the rust products reach the concrete and thus initiate cracking. The second part of the propagation phase, T_{cp}, is referred to

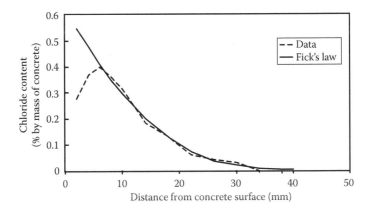

Figure 8.8 Typical chloride content profile fitted to Fick's second law of diffusion.

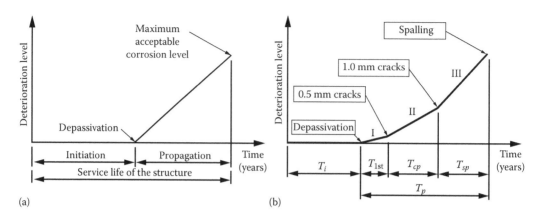

Figure 8.9 Chloride-induced deterioration progression. (a) Two-stage process; (b) four-stage process. (From Kenshel, O., Influence of Spatial Variability on Whole Life Management of Reinforced Concrete Bridges, PhD thesis, Trinity College Dublin, Dublin, Ireland, 2009.)

as the time to crack propagation to a crack size of approximately 1.0 mm. The third part of the propagation phase, T_{sp}, is referred to as the time to severe cracking and spalling.

The time to corrosion initiation, T_i, is commonly derived from Fick's second law of diffusion and is computed by rearranging Equation 8.17 and introducing $x = C_d$ where C_d is the reinforcement cover depth:

$$T_i = \frac{C_d^2}{4D_{app}} \left[\left(erf^{-1} \left(\frac{C_{cr} - C_s}{C_i - C_s} \right) \right)^{-2} \right] \tag{8.18}$$

where C_{cr} is the critical chloride concentration.

Cylindrical layers of concrete are considered around the reinforcing bars. In the literature, two approaches to analytical crack-modelling can be found, illustrated in Figure 8.10:

- The thick-walled uniform cylinder approach (TWUC) where the concrete is modelled as a single layer, which becomes perfectly plastic when in tension (Figure 8.10a)
- The thick-walled double cylinder approach (TWDC), which considers the concrete cylinder as having two parts: a cracked inner cylinder and an uncracked outer one (Figure 8.10b)

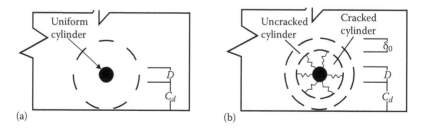

Figure 8.10 Analytical models of corrosion-induced cracking. (a) TWUC; (b) TWDC. (From Reale T. and O'Connor A., *Journal of Construction and Building Materials*, 36, 475–483, 2012.)

Chernin and Val (2011) have examined both models in detail, particularly the theoretical shortcomings with the inherent assumptions. Reale and O'Connor (2012) identified a number of problems with commonly used TWUC models, herein termed the Bazant model (Bazant 1979), the Liu and Weyers model (1998) and the El Maaddawy Model (El Maaddawy and Soudki 2007). They concluded that a modified El Maaddawy model, revised to correct a trigonometric error in estimation of the length of the crack surface, was the most appropriate for predicting the time-to-first-crack:

$$T_{1st} = \left[\frac{7117.5(D+2\delta_0)(1+v+\psi)}{i_{corr}E_{ef}} \right] \left[\frac{(2C_d+D)f_{ct}}{D} + \frac{2\delta_0 E_{ef}}{(D+2\delta_0)(1+v+\psi)} \right] \qquad (8.19)$$

where

T_{1st} is the time (in years) from corrosion initiation to first cracking (defined by the appearance of a hair-size crack on the concrete surface).

D is the diameter of the reinforcing bar.

C_d is the clear concrete cover (mm).

δ_0 is the thickness ($= 12.5 \times 10^{-3}$ mm) of the porous zone around the steel bar, which will have to be filled first before stresses due to rust expansion can be generated.

ψ is a factor dependent on D, C_d and δ_0.

i_{corr} is the corrosion rate density (μA/cm^2), which depends on many factors, such as the availability of oxygen and moisture, and is often obtained from field measurements.

E_{ef} is the effective elastic modulus of the concrete, which is equal to $[E_c/(1 + \phi_{cr})]$.

E_c is the elastic modulus of the concrete (N/mm^2).

ϕ_{cr} ($= 2.0$) is the creep coefficient.

v ($= 0.18$) is the Poisson's ratio for the concrete.

f_{ct} is the tensile strength of the concrete (N/mm^2).

Vu et al. (2005) proposed an empirical model to estimate the period of time it takes for the corrosion-induced crack to propagate from a hair size (typically assumed in the order of 0.05 mm) and reach a maximum size of w_{lim} (mm) (sub-stage II in Figure 8.9b). The model is based on accelerated corrosion tests conducted in the laboratory involving eight RC slabs with 16-mm-diameter reinforcing bars and varying water/cement ratios and cover depths. Assuming a constant corrosion rate, the time it takes for the corrosion-induced crack to propagate from a hair size to the limiting width w_{lim} was estimated as follows:

$$T_{cp} = k_R \frac{0.0114}{i_{corr}} \left[A\left(C_d/wc\right)^B \right] \text{ for } 0.33 \text{ mm} \leq w_{lim} \leq 1.0 \text{ mm} \qquad (8.20)$$

where

$$k_R \approx 0.95 \left[\exp\left(-\frac{0.3i_{corr}^{exp}}{i_{corr}} \right) - \frac{i_{corr}^{exp}}{2500 i_{corr}} + 0.3 \right]$$

In these equations, A and B are empirical constants, k_R is a rate of loading corrosion factor and i_{corr}^{exp} is the accelerated corrosion rate used in the experiment. The developers of the model used an impressed corrosion rate of 100 μA/cm^2 in their testing so $i_{corr}^{exp} = 100$ μA/cm^2. The parameters are $A = 65$ and $B = 0.45$ for $w_{lim} = 0.3$ and $A = 700$ and $B = 0.23$ for $w_{lim} = 1.0$ mm.

The factor k_R was developed to allow for the high rate of loading resulting from the use of high corrosion rates in the accelerated corrosion tests. Mullard and Stewart (2011) further developed these models considering various concrete covers, concrete tensile strenghts, reinforcing bar diameters and the effect of concrete confinement.

As the corrosion process progresses, the cross-sectional area of the reinforcement of an RC member will be reduced, leading to a reduction in the capacity of individual elements and by implication, in the bridge as a whole. If the corrosion is assumed to be uniform (Figure 8.11), the loss of reinforcement diameter can be described using Faraday's law of electrochemical equivalence (Andrade et al. 1993). Pitting corrosion, in contrast to general corrosion, concentrates over small areas of the reinforcement. For further information on modelling the effects of pitting corrosion, the reader is referred to Val and Melchers (1997) and Stewart (2009).

Faraday's law indicates that a constant corrosion rate of 1.0 µA/cm² corresponds to a uniform loss of bar diameter of 0.0232 mm per year (Andrade et al. 1993). If the corrosion rate is assumed to be constant over time, then the remaining cross-sectional area of corroding main reinforcement after t years $A_s(t)$ can thus be estimated as

$$A_s(t) = \sum_{1}^{n_b} \frac{\pi[D_0 - \Delta D(t)]^2}{4} \geq 0 \tag{8.21}$$

where n_b is the total number reinforcing bars, D_0 is the original bar diameter and where the reduction of bar diameter at time, t, is

$$\Delta D(t) = 0.0232 i_{corr} (t - T_i) \tag{8.22}$$

Determination of the rate and extent of loss of steel cross-sectional area can be linked to a reduction in capacity of the bridge element as a function of time. Stochastic modelling of this process can be incorporated directly into a probabilistic analysis to identify how the reliability index, β, varies as a function of time due to the extent of deterioration. Where available, SHM information from the structure can/should be employed to update statistical models of deterioration.

An important consideration in assessing the temporal variation of the β-index is the spatial variability of the level of deterioration across the bridge. Different elements, as a function of the design code, have different levels of inherent durability (i.e. due to varying water/cement ratio, varying cover, etc.) and are also exposed to different concentrations of deleterious agents (e.g. corrosion-inducing chlorides). As a result, levels of deterioration can vary across the structure and even within elements of the structure (e.g. from deck to piers or within the deck). The importance of modelling this spatial variability to improve accuracy, and reduce uncertainty, in the computation of β has been shown by a number of researchers (Engelund and Sorensen 1998; Li et al. 2004; Karimi et al. 2005; Stewart and Mullard 2007; O'Connor and Kenshel 2012).

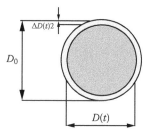

Figure 8.11 General uniform corrosion.

8.6 LOAD MODELLING

When assessing an existing bridge, it is possible to have a more accurate characterisation of the loading than during design. The consequence is that it is often justifiable to reduce the associated partial safety factors for load at the ULS and SLS and in the load combinations to reflect the reduced uncertainty (COST345 2004). Using knowledge of the actual loads also results in a more accurate evaluation of the reliability index in a probabilistic assessment. In performing an assessment, the assessor should take account of all likely loading scenarios. This section discusses the modelling of permanent and quasi-permanent loads (e.g. dead and superimposed dead loads) and variable imposed loads (e.g. traffic loads). Where relevant, modelling of stochastic loading scenarios should include environmental effects such as those due to wind, temperature, and so on. These are, however, not included in this section. The interested reader is rather referred to JCSS (2000a) for detailed treatment of appropriate stochastic modelling.

8.6.1 Permanent and quasi-permanent loads

Permanent and quasi-permanent loads are taken here to include the weight of the structure and earth pressures.

In a probabilistic safety assessment, permanent loads are generally modelled by normally distributed variables (NKB 1978). Here, a distinction is made between the weight of the structure itself, G, and what is sometimes referred to as superimposed dead load, that is, dead load that could be changed such as asphalt and ballast, G_W. In line with DRD (2004), the following can be used as a starting point:

- G is assumed to be normally distributed with a CoV $V_M = 5\%$.
- G_W is assumed to be normally distributed with a CoV $V_M = 10\%$.

The higher level of uncertainty associated with the superimposed dead load is consistent with the rationale behind higher partial factors having been historically applied to this load type. Note, however, that permanent loads from different sources may be assumed to be stochastically independent, and uncertainties can be reduced significantly by measurement.

In addition to the given variations, model uncertainty (V_{I_M}) should also be taken into account to allow for (i) uncertainties in the load calculation model and (ii) uncertainties in the load effect calculation model. The model uncertainty may be incorporated into the analysis by increasing the CoV of the action, V_M, as

$$V = \sqrt{V_M^2 + V_{I_M}^2} \tag{8.23}$$

Thus, if the CoV for self-weight, G, is taken as $V_M = 5\%$ with a CoV for model uncertainty of, $V_{I_M} = 5\%$ (O'Connor and Enevoldsen 2008), then a load effect resulting from G is modelled as normally distributed with a mean value $\mu_G = 1.0$ and a CoV V of 7.1%. Note that this assumes no bias in the estimation of self-weight. For consideration of bias, the reader is referred to Galambos et al. (1982), Ellingwood et al. (1982) or Das (1997).

Other permanent loads that may be considered include

- Soil pressure
- Water pressure
- Differential settlement

- Concrete creep and shrinkage
- Prestress

8.6.2 Variable imposed loads

Of the load effects to be determined in bridge assessment, the most variable are those induced by imposed traffic loads, LL. Traditionally, these effects have been determined in calculations employing deterministic loading models. Deterministic loading models have in the past been derived based on practical experience or more recently in model calibration studies (O'Connor et al. 2001). In both cases, the parameters of the model, traditionally with uniformly distributed and concentrated components or vehicle models, are selected such that they will provide load effects with an acceptably low probability of exceedance. In the calibration studies performed for the normal loading model for Eurocode 1 (EN 1991-2 2003), maximum-in-1000-year load effects were calculated for spans ranging from 5 to 200 m for one to four lanes of traffic and for structural forms ranging from simply supported to multiple continuous spans to fixed–fixed beams (O'Connor et al. 2001). The requirement to be widely applicable results in considerable conservatism in the notional loading models. This conservatism is relatively unimportant in the design of new highway structures when considered from a cost–benefit perspective. However, in the assessment of existing structures, it can be significant, particularly if it leads to unnecessary repair/replacement of a safe structure. In recognition of this, assessment practitioners and researchers have increasingly attempted to take account of site-specific traffic data in the determination of maximum load effects (Ghosn and Moses 1986; Minervino et al. 2004; O'Connor and OBrien 2005; Sivakumar et al. 2007; OBrien and Enright 2011; Enright and OBrien 2013). Detailed discussion of a statistical methodology for probabilistic modelling of traffic loads is presented in Section 2.3.2, and as such, it is not proposed to revisit it here. Application of these methods is presented in Chapter 9.

Model uncertainty in the traffic load considers the uncertainty associated with the definition of load itself and the uncertainty resulting from the transformation of the load to a load effect. According to Nordic and Danish recommendations for probability-based assessment of bridges, the CoV associated with a road traffic load model, V_{I_f}, is taken as 10%, 15% or 20% for a level of uncertainty, considering the loading of small, medium and large, respectively (DRD 2004).

8.7 PROBABILISTIC ASSESSMENT OF LS VIOLATION

Equations 8.3 and 8.4 provide a formulation for evaluation of the exceedance probability at the critical LS, as determined from a deterministic analysis, when that LS is not time-dependent. In assessing highway bridges, loading is typically dominated by the combination of dead and imposed loads, with other loads playing a minor role. Thus, the LS function may often be represented as

$$Z = G(R, S) = R - (G + G_W + LL) \tag{8.24}$$

where the failure state is reached when $Z < 0$.

More complex load combinations consist of finding an equivalent loading system to represent the effect of two or more stochastic load processes acting in combination or individually in a purely additive manner. The reader is referred to the literature for in-depth treatment of load combination modelling (e.g. Melchers 1999).

8.8 COMPONENT VS. SYSTEM RELIABILITY ANALYSIS

A component (or elemental) probabilistic analysis offers insights into the β-index for the individual element and the sensitivities of this β to the modelled parameters (i.e. $\partial \beta / \partial x_i$). Often, however, a system analysis is required to gain an estimate of the overall bridge safety. System reliability concepts arose as an extension to component failure mode analysis for cases where it was apparent that multi-component behaviour had an impact on the true structural behaviour and consequently on the computed β-index. Series and parallel systems are considered as the two fundamental systems. A series system fails if any of its components fail or if a component can fail by different failure modes. A parallel system fails if all of its elements have reached failure. Parallel systems are appropriate in modelling redundant bridges whereby failure of one component does not imply failure of the structure as a whole. Some structural systems consist of combinations of these, whereas others are of a more complex nature, containing conditional aspects. For more information, the reader is referred to the literature (Melchers 1999; Nielson and DesRoches 2007).

Chapter 9

Case studies

9.1 INTRODUCTION

The strength of the probabilistic approaches outlined in Chapter 8 lie in their ability to derive structure-specific safety levels. Thus, for structures that have marginally failed a deterministic assessment, it is sometimes possible to demonstrate adequate capacity at the critical limit state via a probabilistic analysis. The actual safety is compared to the required safety in a structure-specific rather than a general sense. This chapter demonstrates the application of these approaches through the examples of three specific bridge structures. All three marginally failed deterministic assessments, and as such, probabilistic assessments were deemed appropriate.

9.2 REINFORCED CONCRETE BEAM-AND-SLAB DECK

The first example considered is a probability-based assessment of a reinforced concrete beam-and-slab bridge. The structure, illustrated in Figure 9.1, carries two lanes of traffic, is 9.6 m wide and spans 8 m.

The superstructure consists of six parallel longitudinal beams at 1.4 m centres, which carry the load with the slab as T-sections. A diaphragm beam is located in the middle of the span. The presence of an upstand has an influence on the stiffness of the outermost beam. The longitudinal beams have a web width of 0.36 m and a flange thickness of 0.2 m, as illustrated in Figure 9.2. The T-sections have haunches beginning at 0.31 m from the top fibre. The upstand beam has breadth and height of 0.3 and 0.46 m, respectively. The T-beams are reinforced with 26 mm diameter bottom layer main reinforcement placed in two rows, whereas the top layer reinforcement is composed of 14 mm diameter bars. Shear reinforcement in the beams is provided in the form of links. The configuration of the reinforcement is as illustrated in Figure 9.2.

9.2.1 Bridge model

Figure 9.3 presents a schematic model of the structural system of the bridge. The section properties for the edge beam and internal T-beams are calculated based on the geometric sections, as illustrated in Figure 9.4. As recommended in Section 6.5, the section properties for each beam are calculated about its own centroid (as opposed to about the bridge centroid). The properties are listed in Table 9.1.

It is required that the structure be capable of simultaneously carrying a 100 t special transport vehicle with a 50 t normal vehicle in unrestricted passage (i.e. no restrictions on position of the vehicles on the structure or on speed of passage). In the deterministic

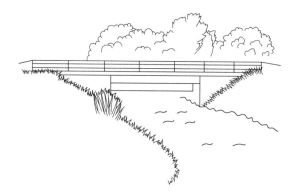

Figure 9.1 Reinforced concrete beam-and-slab bridge.

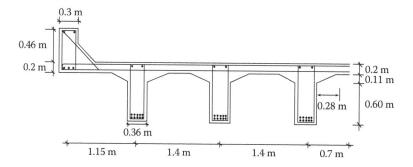

Figure 9.2 Cross section.

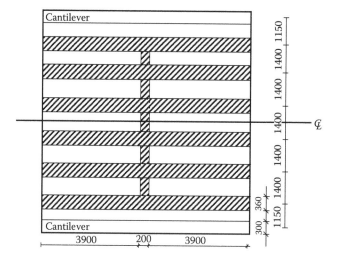

Figure 9.3 Schematic plan of the bridge (dimensions in mm).

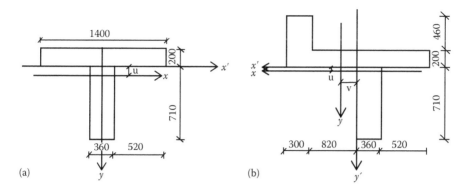

Figure 9.4 Beam section geometries. (a) Internal beams (ignoring haunches). (b) Edge beams (ignoring haunches).

Table 9.1 Section properties

Element	A (m²)	u (m)	v (m)	I_{xx} (m⁴)	I_{yy} (m⁴)
Internal beam	0.54	0.12		4.75×10^{-2}	4.85×10^{-2}
Edge beam	0.79	0.0108	0.2	7.61×10^{-2}	38.0×10^{-2}

assessment, the mid-span moment in the edge beams was found to marginally exceed the capacity at the ultimate limit state (ULS) for this loading scenario. It was therefore considered valuable to perform a probabilistic assessment. In the probabilistic assessment, exceedance of the critical limit state may be described as

$$g \leq 0 \quad \text{where} \quad g = M_{cap} - M_{app} \tag{9.1}$$

where the moment capacity is a function of a number of variables (refer to Figure 9.5):

$$M_{cap} = M_{cap}(h, b_f, b_w, t_f, d_x, d_y, c, A_s, f_c, f_y)$$

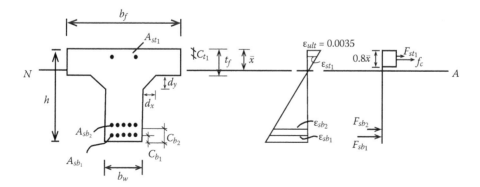

Figure 9.5 Nomenclature. h = overall section height; b_f = flange breadth; b_w = web width; t_f = flange thickness; d_x, d_y = haunch dimensions; c = cover to reinforcement; A_s = reinforcement area; f_c = concrete strength; f_y = steel yield strength.

and

$$M_{app} = M_{DL} + M_{SDL} + M_{LL}$$

where M_{DL} represents the moment due to dead load (DL), M_{SDL} represents the moment due to superimposed dead load (SDL) and M_{LL} represents the moment due to live traffic load (LL).

An influence surface was developed to determine the mid-span moment induced in the edge beam by stochastically generated vehicle loads on the structure. The influence surface was generated using a grillage model of the bridge.

9.2.2 Probabilistic classification and modelling

This section outlines the modelling of the input stochastic variables and deterministic parameters, which were employed to describe the loading on and resistance of the structure. Consideration is given to the required safety level as discussed in Chapter 8.

Required safety level, β_{target}
The requirements at the ULS for structural safety are considered with reference to failure types and failure consequences. For the beam-and-slab bridge, the critical limit state is a ductile failure mode, that is, exceedance of the ULS mid-span moment capacity of the edge beams. With reference to Table 8.5, it is considered appropriate to select *Failure Type II – ductile failure*. The implication is that the safety requirement for the structure at the ULS is $\beta \geq 4.7$.

Modelling of variables relating to capacity
Three strength parameters are modelled as stochastic in the analysis: the concrete compressive strength f_c and the strength of the ordinary reinforcing steel, f_y, for bar diameter ≤ 16 and ≥ 16 mm.

Concrete compressive strength f_c
The concrete compressive strength is determined from tests/drawings from which it was determined that the concrete in the superstructure has a mean strength $\mu_{f_c} = 43.5$ N/mm² with a corresponding coefficient of variation (CoV) V_m of 0.35. As discussed in Section 8.4.5, it is important to consider the model uncertainty in the determination of statistical parameters. Model uncertainty is modelled as a logarithmic normally distributed stochastic variable with an expected value of 1.0, by which the strength variable is multiplied. The accuracy of the calculation model is taken as Normal, the uncertainty of the material properties in construction is classified as Medium, whereas the material identity is classified as Normal. Thus, from Equation 8.16 and Table 8.12

$$V_{I_m} = \sqrt{0.06^2 + 0.06^2 + 0.06^2 + 2((0 \times 0.06) + (0 \times 0.06) + (0 \times 0.06))} \times 0.35 = 10.4\%$$

The CoV including model uncertainty, V, is calculated from Equation 8.23 as

$$V = \sqrt{V_m^2 + V_{I_m}^2} = 36.5\%$$

The resulting standard deviation, including model uncertainty, is calculated as

$$\sigma_{f_c} = \mu_{f_c}V = 43.5 \times 0.365 = 15.88 \text{ N/mm}^2$$

As discussed in Section 8.4.1, the variable is modelled using a lognormal distribution.

Yield strength of the normal reinforcing steel, f_y

The classification of the reinforcing steel, as indicated on the drawings, was equivalent to smooth bar, class Fe 360 (see Table 8.10). The T-beams are reinforced with 26 mm diameter bottom layer main reinforcement, while the top layer reinforcement is 14 mm diameter. From Table 8.10, for ordinary reinforcing, steel for diameter $\leq$ 16 mm and diameter $>$ 16 mm with f_{yk} = 235 and 225 N/mm², respectively, μ_{f_y} = 304 and 293 N/mm² for the top and bottom reinforcement, respectively. The standard deviation associated with non-prestressed reinforcement according to Table 8.10 is 25 N/mm². The model uncertainty is taken into consideration as in the compressive strength of the concrete, which means that (1) the accuracy of the calculation model, (2) the uncertainty for the material resistance in the construction and, finally, (3) material identity are considered. For the original reinforcing steel, the presence of test evidence was taken to reflect a reduction in the uncertainty associated with the reinforcement properties. As such, (1) the accuracy of the calculation model was taken as Normal, (2) the uncertainty for the material resistance in the construction was assumed as Medium and, finally, (3) material identity was taken as Normal.

CoV for steel diameter $\leq$ 16 mm:

$$V_m = \frac{\sigma_{f_c}}{\mu_{f_c}} = 8.22\%$$

$$V_{I_m} = \sqrt{0.06^2 + 0.06^2 + 0.06^2 + 2((0 \times 0.06) + (0 \times 0.06) + (0 \times 0.06)) \times 0.0822} = 10.4\%$$

$$V = \sqrt{V_m^2 + V_{I_m}^2} = 13.25\%$$

CoV for steel diameter $>$ 16 mm:

$$V_m = \frac{\sigma_{f_c}}{\mu_{f_c}} = 8.53\%$$

$$V_{I_m} = \sqrt{0.06^2 + 0.06^2 + 0.06^2 + 2((0 \times 0.06) + (0 \times 0.06) + (0 \times 0.06)) \times 0.0853} = 10.4\%$$

$$V = \sqrt{V_m^2 + V_{I_m}^2} = 13.45\%$$

These variation coefficients are equivalent to standard deviations of 40.29 and 39.40 N/mm² for the diameter $\leq$ 16 mm and $>$ 16 mm reinforcement, respectively. As discussed in Section 8.4.1, the variable is modelled using a lognormal distribution.

Other resistance variables
The elastic modulus of the materials is modelled as a deterministic parameter with its value for concrete and steel being taken as 35,000 and 200,000 N/mm², respectively. The parameters describing the cross section of the T-beam under investigation – (1) section height, (2) flange width, (3) web width, (4) flange thickness, (5) haunch width and (6) haunch height – were also modelled deterministically. The main longitudinal steel in the T-beams is provided by the ordinary reinforcement. On the bottom of the beam, 10 No. 26 mm diameter bars in two layers provide a reinforcement area of 5309 mm². In the top of the beam, 2 No. 14 mm diameter bars in a single layer provide a reinforcement area of 308 mm². The cover to the top and bottom reinforcement layers is modelled deterministically. The bottom steel is modelled in two layers of 5 × 26 mm diameter bars with the bottommost layer having cover of 51 mm and the second layer having cover of 103 mm. The top steel consists of a single layer of 14 mm diameter bars at 45 mm cover.

Modelling of variables relating to loading
Self-weight of the structure
The load effects induced by the self-weight of the structure are a superposition of the effects of the beams, slab and pavement. The magnitude of the load effect induced is modelled as stochastic (see Section 8.6.1). For the DL, the mean value is taken as 1.0 with a CoV including model uncertainty of 7.07% (i.e. $V_m = 5.0\%$, $V_{I_m} = 5.0\%$; therefore, $V = \sqrt{V_m^2 + V_{I_m}^2} = 7.07\%$ – see Section 8.6.1). For the surfacing, the mean value is taken as 1.0 with a CoV, including model uncertainty, of 11.18% (i.e. $V_m = 10.0\%$, $V_{I_m} = 5.0\%$; therefore $V = \sqrt{V_m^2 + V_{I_m}^2} = 11.18\%$).

Imposed loading
Of the loads to be modelled on a highway bridge, by far, the most variable are those due to traffic loading, *LL*. This variability results not only from the stochastic variables describing the individual vehicles themselves – weight, axle spacing, speed, dynamic amplification etc. but also from the probability of multiple presence longitudinally within an individual lane or transversely in multiple lanes. As discussed in Section 2.3.2, the critical loading events for two-lane, two-direction bridges with influence length up to approximately 50 m occur due to (1) meeting events between ordinary trucks and (2) meeting events involving heavy transports (e.g. cranes, low-loaders or permit vehicles) with ordinary trucks. In both cases, the extreme distribution function of the load effects may, for example, be obtained from the so-called thinned Poisson process (Ditlevsen 1994; Ditlevsen and Madsen 1994) only arrival and meeting events including the heaviest groups of trucks in the various traffic situations are considered. Figure 9.6 indicates one such meeting event.

For a two-lane bridge, the extreme distribution F_{max} of the considered loading event q can be obtained from

$$F_{max}(q) = \exp[-(\nu_1 - \nu_{12})T(1 - F_1(q))]\exp[-(\nu_2 - \nu_{12})T(1 - F_2(q))]$$
$$\times \exp[-\nu_{12}T(1 - F_{12}(q))] \tag{9.2}$$

where ν_1 and ν_2 are the intensities of the considered traffic (e.g. number of vehicles of type considered per day) in lanes 1 and 2, respectively and ν_{12} is the intensity of the meeting

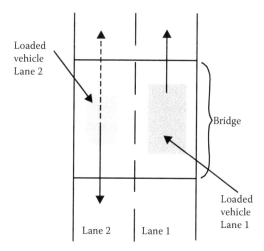

Figure 9.6 Typical uni-directional or bi-directional meeting event.

event T is the considered reference period for the extreme distribution (e.g. one traffic year).

The distribution for the load event in lane 1, $F_1(q)$, and lane 2, $F_2(q)$, and the distribution of load due to simultaneous LL in both lanes, $F_{12}(q)$, must be determined (O'Connor and Enevoldsen 2008). These three distributions do in general include modelling of (1) the number, configuration and weights of trucks, (2) the longitudinal and transverse occurrence in bridge lanes and (3) the dynamic amplification of the static truck load.

Number, configuration and weight of trucks
Table 9.2 indicates the statistics provided for the expected annual frequency of special heavy vehicles of various classes (>100 t) for various route types (DRD 2004).

The configuration of the 100 t vehicle types is as illustrated in Figure 9.7a. For ordinary six axle vehicles, the axle configuration illustrated in Figure 9.7b is employed (DRD 2004;

Table 9.2 Annual frequency of standard heavy transports

Route/class	100	125	150	175	200
Motorway	100	50	50	50	50
Primary	50	20	20	20	20
Secondary	20	10	10	10	10

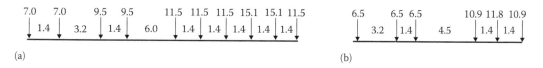

(a)   (b)

Figure 9.7 Vehicle silhouettes (axle spacings in m, axle weights in tonnes). (a) 100 t vehicle in lane 1. (b) 50 t vehicle in lane 2.

O'Connor and Enevoldsen 2008). The expected annual frequency of ordinary transports may be determined from weigh-in-motion (WIM) records or traffic counts at the location of the bridge. For the structure considered, the expected frequency was taken as 100/year and 1000/day for the 100 and 50 t vehicles, respectively. For each truck class, the mean gross vehicle weight is as specified (i.e. here for the considered loading case, as 100 and 50 t, respectively), whereas the standard deviation on the weight is taken as 5 t for all classes (DRD 2004; O'Connor and Enevoldsen 2008).

Longitudinal and transverse occurrence in bridge lanes
For the analysis performed, the transverse location of the vehicles is modelled as a truncated normally distributed random variable with the mean position taken as the centre of the driving lane and a standard deviation of 0.24 m (DRD 2004; O'Connor and Enevoldsen 2008). The truncation in the distribution is necessary to reflect the physical dimensions of the lane. Longitudinally, the vehicles are generally located at the critical influence ordinate. This represents a conservative assumption.

Dynamic amplification of static truck load
In the analysis, the conservatism of ignoring the inverse relationship between vehicle weight and dynamic amplification factor should be considered. The dynamic factor is therefore modelled as (DRD 2004)

$$K_s = 1 + \varepsilon \tag{9.3}$$

where ε is the dynamic increment, which for vehicles in normal passage is modelled as two independent, normally distributed stochastic variables N(41.5/W; 41.5/W) for an influence length $l > 2.5$ m and as $\mu = (83/W) - (16.6/W)l$; $\sigma = \mu$ for $l < 2.5$ m, where W is the total vehicle weight in kilonewtons (kN). Modelling the dynamic amplification factor implicitly assumes (1) an inverse proportionality between the dynamic amplification and vehicle weight (see Figure 2.16) and (2) a reduction in the CoV with increasing weight, which compares well with the literature (Hwang and Nowak 1991; Kirkegaard et al. 1997; OBrien et al. 2010; González et al. 2011; Caprani et al. 2012).

The expected value for model uncertainty in the LL is defined in terms of a judgement factor, I_f, which is assumed to be normally distributed with a mean value of 1.0 and a CoV V_{I_f} taken as 10%, 15% and 20% for a level of uncertainty assumed as small, medium and large, respectively – see Section 8.6.2 (DRD 2004; O'Connor and Enevoldsen 2008).

Additional parameters
The special transport (i.e. 100 t vehicle) is assumed to be driving at a speed of 60 km/h. The speed for the ordinary trucks is conservatively modelled as 80 km/h (O'Connor and Enevoldsen 2008). The vehicle length is modelled according to the length of the standard vehicles in Figure 9.7 as 19.0 m for the 100 t and 11.9 m for 50 t vehicles, respectively (DRD 2004). The bridge influence length is taken as 8.0 m. This value is equivalent to the total bridge length. The influence length is defined as the length of the structure that, when loaded, contributes to the magnitude of the load effects at a specified location (where relieving zones are excluded from this length). The duration over which trucks are assumed to use the road network is assumed to be 15 h/day (O'Connor and Enevoldsen 2008).

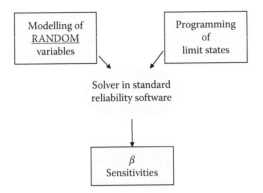

Figure 9.8 Algorithm for analysis.

9.2.3 Results of probabilistic assessment

Determination of the probability of failure P_f is performed using the FORM technique discussed in Chapter 8 by modelling the stochastic variables of load and resistance at the critical limit state as outlined above. A variety of software packages exist to perform this type of analysis (e.g. OpenSees, Proban, STRUREL). Schematically, the process of analysis is as illustrated in Figure 9.8.

The result is a computed probability of failure $P_f = 3.48 \times 10^{-7}$ or $\beta_{FORM} = 4.96$. As the computed β exceeds the target level of 4.7, it may be concluded that the structure has adequate safety at the considered limit state for the loading scenario considered (i.e. simultaneous presence of 100 and 50 t vehicles on the structure). It is important to consider not only the output β value from the analysis but also (1) the relative importance of modelling the uncertainty of the considered random variables, (2) the sensitivity of the computed β value to changes in the modelled variables and (3) the values of the variables at the β point.

Importance factors

The importance factors allow identification of the random variables, which have the greatest impact on the final computed reliability index β (Madsen et al. 1986). These are the variables that will be most beneficial to focus on collecting additional information to reduce their uncertainty. Conversely, for those that demonstrate small importance, it may be possible to replace the variable by a deterministic value, for example, the median value. Table 9.3 presents the importance factors computed for this example.

Table 9.3 Importance factors

Variable	Importance (%)
Steel strength (dia > 16 mm)	74.8
Model uncertainty, traffic load	13.9
Transverse location of 100 t vehicle	6.5
Dynamic amplification of 100 t vehicle	2.1
Dead load	0.9
Superimposed dead load	0.8
Others	1.0

From these results, it is apparent that the variable of greatest importance with respect to the computed β is the steel strength f_y of the beam soffit reinforcement, that is steel whose diameter exceeds 16 mm. The model uncertainty associated with the LL model and the transverse location of the 100 t vehicle are also of importance but to a far lesser degree. The other variables, for example, DL and SDL, are seen to have little significance.

Sensitivity analysis

A sensitivity analysis should also be performed to assess the influence of a change in the modelled parameters on the computed β index. Here, the sensitivity of β to a 10% change in the various parameters (i.e. mean and standard deviation) is assessed. The most significant of these parameters are presented in Table 9.4.

It is demonstrated that a +10% change in the mean strength of the reinforcing steel with diameter > 16 mm, i.e. an increase from 293 to 322 N/mm², increases the computed β by 1.02, from 4.96 to 5.98. On the other hand, an increase in the standard deviation by 10%, from 39.4 to 43.3 N/mm², reduces the computed β by −0.39, from 4.96 to 4.57. This result is of significance as the revised β is less than the target value of 4.75. Hence, it may be considered worthwhile to collect more information on this reinforcement in an attempt to improve the accuracy of its statistical characterisation.

Values of the variables at the design point

Reviewing the values of the variables at the β point is important to understand the extent to which certain variables have moved from their mean value into the tail of the distribution as opposed to those who have moved little, if at all, from the mean. The values of the variables at the β point should reflect the results outlined in Tables 9.3 and 9.4.

Table 9.4 Sensitivity analysis

Variable	Name	Distribution	$\Delta\beta$ for a +10% change in mean	$\Delta\beta$ for a +10% change in cov
Concrete strength	f_c	Lognormal	0.00	0.00
Steel strength, dia > 16 mm	f_{y1}	Lognormal	1.02	−0.39
Steel strength, dia ≤ 16 mm	f_{y2}	Lognormal	0.00	0.00
Model uncertainty, traffic load	ModUnc	Normal	−0.37	−0.05
Transverse location of 100 t vehicle	TR1	Truncated normal	0.03	0.00
Dead load	DL	Normal	−0.14	0.00
Superimposed dead load	SDL	Normal	−0.08	0.00

Note: Change of β for a 10% change in the mean value and CoV of the modelled parameters.

Table 9.5 Values of variables at β point

Variable	Name	Distribution	Mean, μ	Standard deviation, σ	Value of variable at β point, x_i^*
Concrete strength	f_c	Lognormal	43.5 N/mm²	15.67 N/mm²	39.7 N/mm²
Steel strength, dia > 16 mm	f_{y1}	Lognormal	293 N/mm²	39.4 N/mm²	159.9 N/mm²
Steel strength, dia ≤ 16 mm	f_{y2}	Lognormal	304 N/mm²	40.2 N/mm²	302 N/mm²
Model uncertainty, traffic load	ModUnc	Normal	1.0	0.1	1.15
Transverse location of 100 t vehicle	TR1	Truncated normal	0.5 m	0.24 m	0.35 m
Dead load	DL	Normal	1.0	0.071	1.03
Superimposed dead load	SDL	Normal	1.0	0.112	1.04

As expected, it is apparent from Table 9.5 that the variable that has moved most from its mean is the steel strength for the reinforcement with diameter > 16 mm. At the β point, this variable has a value of 159.9 N/mm² compared to its mean value of 293 N/mm².

9.3 POST-TENSIONED CONCRETE SLAB DECK

The next case considered is the probability-based assessment of a prestressed post-tensioned slab bridge (O'Connor and Enevoldsen 2008). The bridge is skewed at 27° with three spans of 6.39, 17.72 and 6.39 m, giving a total length of 33.88 m (including side span overhangs of 1.69 m). The structure carries four lanes of traffic over its 28 m width. Transversely, the bridge is supported on 10 columns at 3.24 m centres (Figure 9.9).

The main structure consists of a 0.5 m deep longitudinally post-tensioned solid slab, which is continuous over the column supports. The bridge elevation and cross section are illustrated in Figures 9.10 and 9.11. A plan of the structure is provided in Figure 9.12. Post-tensioning is provided in groups of eight tendons, with each tendon composed of 12 No. 7 mm diameter strands. The tendons are profiled longitudinally. Mild steel reinforcement is provided longitudinally and transversely; shear reinforcement is not provided in the slab.

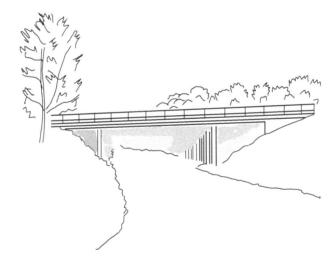

Figure 9.9 Post-tensioned concrete slab bridge.

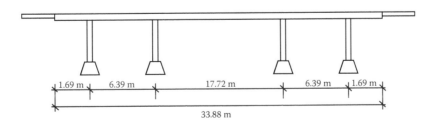

Figure 9.10 Bridge elevation.

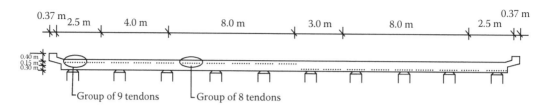

Figure 9.11 Bridge cross section.

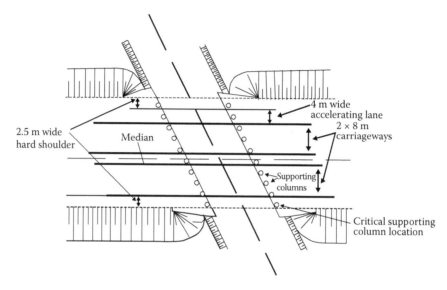

Figure 9.12 Bridge plan.

9.3.1 Bridge model

A plate FE model of the structure was developed with the width of modelled deck corresponding to that supported by six columns (see Figure 9.13). This width of the structure was reduced to optimise solver time for the FE model. It was based upon the required width for positioning of the vehicles and adequate distribution of the sectional forces. The slab

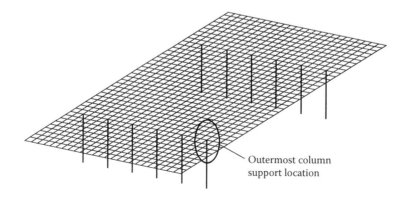

Figure 9.13 Finite element model of the bridge.

stiffness was modelled as orthotropic, and as longitudinal stresses dominate, the elastic modulus in the transverse direction is taken conservatively as, $E_y = 1/5\ E_x$.

As with the previous example, it is required that the structure should be capable of simultaneously carrying a 100 t special transport vehicle with a 50 t normal vehicle in unrestricted passage. Deterministic assessment considered (1) plastic bending capacity assessment of the slab at the ULS, where assessment of the foundation forces and bearing capacity was also performed; (2) shear capacity of the slab at ULS, including punching shear assessment; and (3) standard SLS checks on deflection, steel stresses, crack widths and so on. While the bridge passed deterministic assessment for cases 2 and 3 for the desired load rating, for case 1, it was found to have insufficient hogging moment capacity at the outermost column–slab intersection location, as illustrated in Figure 9.13. At this location, the structure passes a deterministic assessment for simultaneous presence of an 80 and a 50 t vehicle but fails with the 100 and 50 t combination. As such, it is considered worthwhile to perform a probabilistic assessment to investigate if the higher load rating could be achieved.

The critical limit state is considered to be violated by exceedance of the ULS slab hogging moment capacity at the location of the outermost column support. Mathematically, exceedance of this limit state may be described as

$$g \leq 0 \quad \text{where} \quad g = M_{cap} - M_{app} \tag{9.4}$$

where the moment capacity is a function of variables related to concrete and steel strength:

$$M_{cap} = M_{cap}(f_c, f_{ps})$$

with f_c = concrete strength and f_{ps} = prestressing steel guaranteed strength and

$$M_{app} = M_{DL} + M_{SDL} + M_{LL} + M_{par}$$

where M_{DL}, M_{SDL}, M_{LL} and M_{par} are moments due to DL, SDL, LL and parasitic moment due to prestress, respectively. The magnitude of M_{LL} is a function of randomly generated vehicle weights and positions on the structure. To facilitate evaluation of M_{LL}, an influence surface was employed.

To obtain values of M_{cap} in the probabilistic analysis, for randomly generated values of the modelled stochastic variables f_c and f_{ps}, a response surface is required. A response surface is a closed form and differentiable limit state surface constructed using a polynomial or other suitable function fitted to the results obtained from a limited number of discrete numerical analyses (Melchers 1999; O'Connor and Enevoldsen 2008). Training of the response surface is discussed in Section 9.3.4, following description of the statistical modelling of the governing variables.

9.3.2 Probabilistic classification and modelling

Required safety level, β_{target}

For the post-tensioned slab bridge deck of the considered structure, the critical limit state is a ductile failure mode. With reference to Table 8.5, it is considered appropriate therefore to select *Failure Type II – Ductile failure*. The implication is that the safety requirement for the structure at the ULS is $\beta \geq 4.7$.

Modelling of variables relating to capacity

Two strength parameters are modelled as stochastic in the analysis:

1. The concrete compressive strength f_c
2. The prestressing steel guaranteed strength f_{ps}

Concrete compressive strength f_c

The concrete in the superstructure has a mean strength, $\mu_{f_c} = 29.4$ N/mm², with a corresponding CoV V_m of 0.18. As in the previous example, the model uncertainty is modelled as a logarithmic normal distributed stochastic variable with an expected value of 1.0. The accuracy of the calculation model is taken as Normal, the uncertainty of the material properties in construction is classified as Medium, whereas the material identity is classified as Normal. Thus, the CoV, including model uncertainty, is $V = 20.8\%$. The resulting standard deviation, including model uncertainty, is 6.11 N/mm².

Yield strength of the prestressing steel f_{ps}

Data regarding the yield strength of the prestressing steel for the bridge is not available. Information contained on the drawings indicates that 7 mm strands were used. Data regarding the guaranteed strength of such tendons estimated at the 0.1% fractile indicate a value of 1422 N/mm² as appropriate for yield, with 1618 N/mm² at ULS, whereas the variation coefficients on the strands are taken as 4% (O'Connor and Enevoldsen 2008). When the variable is modelled using a lognormal distribution, this corresponds to a mean value of 1490 N/mm² and a standard deviation of 59.6 N/mm². The model uncertainty is taken into consideration as in the compressive strength of the concrete: (1) the accuracy of the calculation model was taken as Normal, (2) the uncertainty for the material resistance in the construction was assumed as Medium and, finally, (3) material identity was taken as Normal. The resulting variation coefficient, including model uncertainty, is $V = 11.14\%$. This is equivalent to a standard deviation of 165.9 N/mm².

Structure response surface

The ULS bending capacity of the post-tensioned slab may be assessed using one of the proprietary software tools available for this purpose. From such an analysis, it was determined that the slab capacity at the critical location – the outermost column support location – is governed by a tributary width of slab at the critical location of 3 m. At this width, the influence of the tendons directly over the support is felt, in addition to a proportion of the tendons (3 of 8) between columns, as illustrated in Figure 9.14.

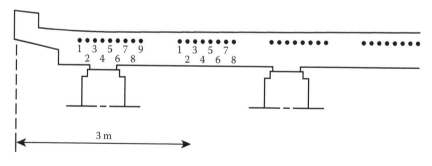

Figure 9.14 Tendons considered in capacity assessment.

Table 9.6 Values of basic resistance variables
employed to train the response surface

f_c (N/mm²)	f_{ps} (N/mm²)
μ	μ
μ – 2σ	0.75μ
μ – 4σ	0.50μ
	0.25μ

Probabilistic load carrying capacity assessment requires that structural capacity be cal-culated for any combination of the basic resistance variables. For this structure, it was necessary to construct a response surface trained upon computed capacity results, evalu-ated using proprietary software, for limited combinations of the basic resistance variables. The various values of these variables employed in training the response surface are out-lined in Table 9.6. The values are selected on the basis of possible values of the variables at the design point. All possible permutations of the values listed in Table 9.6 are selected in training the response surface. It is important to point out that the trained response surface should be thoroughly tested, to validate its accuracy, before employing it in subsequent analysis.

Other resistance variables
The elastic modulus of each material is modelled as a deterministic parameter with the values for concrete and steel being taken as 35,000 and 200,000 N/mm², respectively. The parameter describing the cross section of the slab under investigation – slab depth – was modelled deterministically. Ordinary (i.e. non-prestress) longitudinal and transverse rein-forcement is also provided. The influence of the ordinary reinforcement on the capacity of the slab is minimal, and as such, they are modelled as deterministic parameters. The cover to the top and bottom reinforcement layers is also modelled deterministically.

Modelling of stochastic variables – Loading
Self-weight of the structure
As in the previous example, for the DL, the mean value is taken as 1.0 with a CoV including model uncertainty of 7.07%. For the surfacing, the mean value is taken as 1.0 with a CoV including model uncertainty of 11.18%.

Parasitic moments
The parasitic moment induced by the prestress is modelled stochastically. As for DL, the mean value is taken as 1.0 with a CoV including model uncertainty of 7.07% (i.e. $V_m = 5.0\%$, $V_{I_m} = 5.0\%$; therefore, $V = \sqrt{V_m^2 + V_{I_m}^2} = 7.07\%$) (O'Connor and Enevoldsen 2008).

Traffic loading
The LL is modelled as outlined in Section 9.2.2.

9.3.3 Results of probabilistic assessment

The probability of failure P_f is found using the FORM technique. The result is a computed probability of failure $P_f = 2.12 \times 10^{-7}$ or $\beta_{FORM} = 5.06$. As $\beta > \beta_{target}$, it may be concluded that the structure has adequate safety at the considered limit state for the required load rating (i.e. simultaneous presence of 100 and 50 t trucks).

Importance factors

The significance of consideration of the output importance factors has been discussed previously. Table 9.7 presents the importance factors for this example.

It is apparent that the strength of the concrete, the LL model uncertainty and the strength of the prestressing steel, with a cumulative importance of >90%, are controlling. It is apparent that the other variables, for example, DL, SDL and so on, are of minor influence only.

Sensitivity analysis

The results of the sensitivity analysis are presented in Table 9.8. It is demonstrated that a +10% change in the mean strength of the concrete, i.e. increase from 29.4 to 32.3 N/mm², increases the computed β by 0.51, that is, from 5.06 to 5.57. An increase in the standard deviation by 10%, from 5.85 to 6.44 N/mm², reduces the computed β by -0.20, from 5.06 to 4.86. Similar analyses can be performed to assess the sensitivity of β to the remaining modelled variables and consequently the robustness of the safety rating.

Table 9.7 Importance factors

Variable	Importance (%)
Concrete strength	39.5
Model uncertainty, traffic load	27.3
Prestressing strength	23.9
Dead load	2.9
Dynamic amplification of 100 t vehicle	1.8
Superimposed dead load	1.5
Transverse location of 100 t vehicle	0.8
Others	2.3

Table 9.8 Sensitivity analysis

Variable	Name	Distribution	$\Delta\beta$ for a +10% change in mean	$\Delta\beta$ for a +10% change in CoV
Concrete strength	f_c	Lognormal	0.51	−0.20
Prestress strength	f_{ps}	Lognormal	0.56	−0.12
Model uncertainty, traffic load	ModUnc	Normal	−0.34	−0.14
Transverse position lane 1	TR1	Truncated normal	0.02	0.01
Dead load	DL	Normal	−0.25	−0.02
Superimposed dead load	SDL	Normal	−0.12	−0.01

Note: Change of β for a 10% change in the value of the mean value and CoV of the modelled parameters.

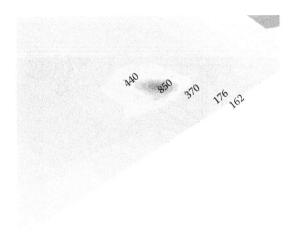

Figure 9.15 M_x from FE model for design point loading.

Values of the variables at the design point

As discussed in Section 9.2.3, the results of this analysis are highly dependent on the accuracy of the influence and response surfaces employed in the analysis. It is therefore necessary to check the validity of the estimates at the design point.

Figure 9.15 presents the values of M_x determined from the finite element (FE) model where the critical loading scenario at the design point is employed in the analysis. At the design point, the total load of the 100 and 50 t vehicles are 1699 and 826 kN, respectively, where these values incorporate dynamic amplification and model uncertainty. The average value for M_x distributed over a 3 m width is calculated from Figure 9.15 as approximately 400 kNm. The value determined by the 'averaged' influence surface at the design point was 406 kNm, representing an acceptable difference of 1.6%.

The values of concrete and prestress reinforcement strength at the design point are 15.8 and 1124 N/mm², respectively. A capacity analysis performed using proprietary software employing these values as input yields a section capacity of 1955 kNm for a 3 m width or 651 kNm/m. This is compared with the response surface estimate of 635 kNm, representing an acceptable difference of 2.6%.

9.4 STEEL TRUSS BRIDGE

The final example considers a probability-based assessment of a steel truss railway bridge (O'Connor et al. 2009). The superstructure of the bridge is composed of riveted trusses with spans of 42.0, 84.0 and 42.0 m, as illustrated in Figure 9.16. Simply supported side approach spans of 22.5 and 11.6 m give a total bridge length of 202.1 m. The main structure is supported at four longitudinal locations, one-fixed plus three-roller (i.e. guided sliding) bearings, which results in the bridge working as a three-span continuous beam.

The primary elements of the superstructure forming the top and bottom chords, the vertical and diagonal truss elements, are hereinafter labelled as Over (O), Under (U), Vertical (V) and Diagonal (D), respectively. The sleepers and rails are supported by a system of secondary longitudinal beams (or floor beams), termed SLB, which in themselves are supported by transverse beams, TB, at 7.0 m centres. The superstructure is therefore subdivided into

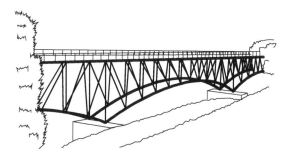

Figure 9.16 Steel truss railway bridge.

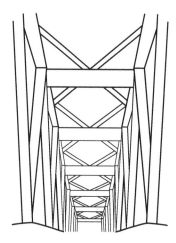

Figure 9.17 View from beneath bridge, showing diagonal wind bracing.

6 + 12 + 6 cells at 7.0 m centres. Between the O-elements, SLBs and TBs, is a system of diamond-formed wind diagonals in a horizontal plane, as illustrated in Figure 9.17. These elements secure the bridge's stability with regards to wind and other transverse loads while also providing capacity for the global effects.

9.4.1 Bridge model

Structural analysis was performed in two phases using a three-dimensional finite element (3D FE) beam model of the bridge, as illustrated in Figure 9.18a. The FE beam model was calibrated using shell and brick element models constructed for specific critical locations (e.g. interaction of the primary and secondary deck elements), as shown in Figure 9.18b. These provided information on appropriate joint stiffness values that were subsequently used in the beam model.

A deterministic assessment of the bridge was performed according to the Swedish Assessment Code, BVH 583.11.2, using the train load model BV-3 (i.e. a train load model with 25 t axles and an 8 t/m line load). The results of this assessment demonstrated that, although the structure had sufficient capacity with respect to SLS and fatigue limit state requirements, at a number of locations, the structure failed to demonstrate the necessary ULS capacity. As such, significant strengthening or complete replacement was deemed necessary by the deterministic assessment. As this conclusion would prove extremely costly, it

Figure 9.18 Finite element modelling of the bridge. (a) 3D FE beam element model. (b) Shell/brick element model.

was considered appropriate to perform a probabilistic assessment of the structure for the critical elements/joints.

With reference to Figure 9.19, the critical limit states to be considered in the probabilistic assessment were identified as (1) the ULS capacity of elements U_7, U_8, SLB (pos 7) and TB (pos 9) and (2) the ULS capacity of the riveted connections 6-U_6, 7-U_6, 7-U_7, 8-U_7, 8-U_8, 7-V_7, 2-D_2, 3-D_2, 3-D_3 and 4-D_3. In this example, the methodology and results are presented for one element only: element U_7.

In the probabilistic assessment, exceedance of the critical limit state for this element may be described as

$$g \leq 0 \text{ where } g = f_y - |\sigma| \tag{9.5}$$

where f_y describes the yield strength of the structural steel and $|\sigma|$ is the induced stress due to applied loads $= \sigma_{F_x} + \sigma_{M_y} + \sigma_{M_z}$ i.e. the sum of axial stress, σ_{F_x} and bending stresses about the y and z axes, σ_{M_y} and σ_{M_z}.

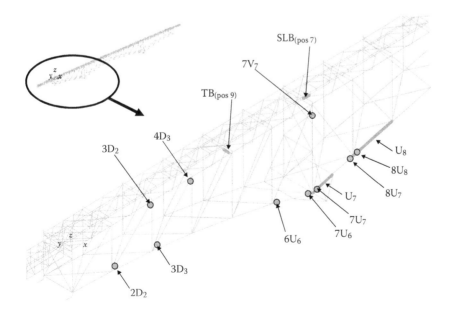

Figure 9.19 Critical elements and joints. (From O'Connor, A. et al., *Structural Engineering International*, 19(4), 375–383, 2009.)

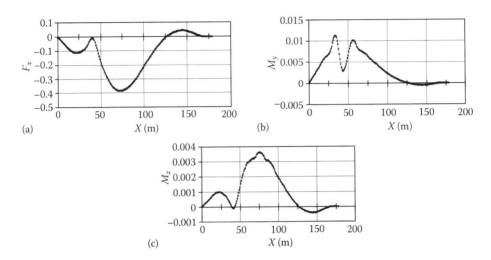

Figure 9.20 Influence lines for element U7. (a) F_x. (b) M_y. (c) M_z. (From O'Connor, A. et al., *Structural Engineering International*, 19(4), 375–383, 2009.)

The element considered for analysis labelled U_7 had a utilisation ratio greater than 1.0 following the deterministic assessment. As a result, it failed the deterministic assessment (utilisation < 1.0 implies the element has sufficient capacity to sustain its loads). In further analysis, it was found that 68% of this utilisation ratio was due to the effects of axial force, F_x, with 31% due to primary bending M_y and just 1% due to secondary bending M_z ($\sigma = \sigma_{F_x} + \sigma_{M_y} + \sigma_{M_z} = 149.9 + 67.9 + 3.3 = 221.1$ N/mm^2, utilisation ratio = 221.1/201 = 1.1 where 201 N/mm^2 is the yield stress of the steel).

From consideration of the influence lines (Figure 9.20), it is apparent that the utilisation ratio for the element is fully controlled by global effects. On this basis, the adverse length to be loaded for the influence lines is approximately 126 m, ignoring the relieving length. The significance of this in the context of probabilistic modelling of train loading will be discussed in Section 9.4.2.

9.4.2 Probabilistic classification and modelling

Required safety level, β_{target}
As for the other examples, the critical limit state is a ductile failure mode so, with reference to Table 8.5, *Failure Type II – Ductile failure* is selected, giving a required safety level at the ULS of $\beta \geq 4.7$.

Modelling of variables relating to capacity
Only one strength parameter is modelled as stochastic in this analysis, the yield strength of the structural steel f_y. Tests performed on the steel provided a mean strength of 331.9 N/mm^2 with a standard deviation of 26.6 N/mm^2 (CoV 8.0%). Model uncertainty is modelled as a logarithmic normal distributed stochastic variable with an expected value of 1.0, which is multiplied by the strength variable. The accuracy of the calculation model is taken as Good, the uncertainty of the material properties in construction is classified as Low (due to steel being the construction material) and the material identity is classified as Good due to the availability of test data. Based on these assumptions, the steel strength is modelled as lognormally distributed with a mean value of 332 N/mm^2. The standard deviation, including model uncertainty, is calculated as in previous examples and found to be 28.93 N/mm^2 (CoV 8.72%).

Modelling of stochastic variables – Loading
Self-weight of the structure
The sectional force induced is modelled as stochastic; see Section 8.6.1. For the DL, the mean value is taken as 1.0 with a CoV, including model uncertainty, of 7.07%. For the SDL (i.e. sleepers, rails, railings, etc.), the mean value is taken as 1.0 with a CoV, including model uncertainty, of 11.18%.

Traffic loading
Stochastic modelling of the train load is based upon 1 month of WIM measurements.

Figure 9.21a presents a histogram of the 28,801 wagon loads (>60 t) recorded. The second mode of the distribution presented in Figure 9.21a is located at approximately 100 t, corresponding to the upper limit of 25 t per axle for a four-axle wagon.

Figure 9.21b presents the tail of the distribution – the frequencies for wagon weights ≥ 104 t. There are 81 occurrences of these overloads in the 1 month of records.

Extreme value modelling of LLs was discussed in Section 2.3. In deriving the extreme value distribution (EVD) for train wagon weights to be used in the subsequent probabilistic analysis, it is important to consider the influence lines for the elements under consideration (i.e. contribution of local and global effects to the overall capacity utilisation). While in some cases it may be appropriate to derive the parameters of the EVD based upon the heaviest single wagon (i.e. for local effects), in many others, it may be more appropriate to consider the heaviest group per train of, for example, 5 or 10 wagons (i.e. for global effects). In the case of element U_7, global effects govern as can be seen from the influence lines presented in Figure 9.20. Hence, the EVD is derived here based upon a group of wagons. As discussed, the adverse length to be loaded for these influence lines is approximately 126 m. The length of wagons considered is 12.5 m. Hence, a group of 10 wagons will be considered in the extreme value analysis.

Figure 9.22a presents the cumulative distribution function (CDF) of the Gumbel EVD fitted to the empirical data for the heaviest group of 10 wagons per train. The data can be seen to fit well to the EVD of the Gumbel form. The equation of the CDF of the Gumbel EVD is

$$F_X(x) = \exp\left[-\exp\left(\frac{-(x-\lambda)}{\delta}\right)\right] \quad -\infty < x < \infty \quad \delta > 0 \tag{9.6}$$

The threshold, λ, and scaling parameter, δ, of the Gumbel cumulative probability, $F_X(x)$, are estimated using a maximum likelihood approach (Castillo 1988).

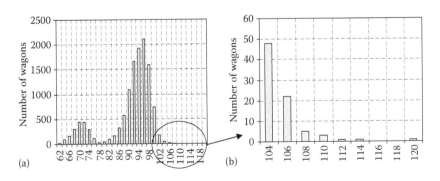

Figure 9.21 Histogram of wagon weights (a) >60 t and (b) ≥104 t. (a) Histogram of wagon weights. (b) Tail of histogram. (From O'Connor, A. et al., *Structural Engineering International*, 19(4), 375–383, 2009.)

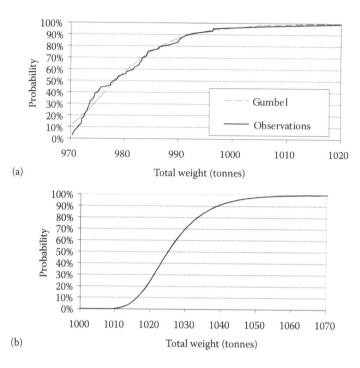

Figure 9.22 EVD based on weights of 10 wagons. (a) Recorded data and fitted distribution. (b) Distribution of annual maxima. (From O'Connor, A. et al., *Structural Engineering International*, 19(4), 375–383, 2009.)

The distribution of the 50 maximum observations $F_X(x)$ (within 1 month) are converted to the distribution of annual maxima $F_{max}(x)$, illustrated in Figure 9.22b, as

$$F_{max}(x) = \exp[-N(1 - F_X(x))] \tag{9.7}$$

with $N = 600$, that is, 12×50, being the annual number of maxima (O'Connor et al. 2009). The distribution parameters derived in this manner for a varying number of wagon groups chosen to model the extreme are listed in Table 9.9, where the values in brackets indicate the mean per wagon, presented for the purpose of comparison.

As expected the mean, μ, per wagon reduces as the number of wagons considered is increased. The CoV is also seen to reduce with an increase in the number of wagons. This is not unexpected given the strict loading protocols employed in configuring/loading train wagons and the expected reduction in variation for the sum of a number of random variables when compared to a single variable.

Table 9.9 Parameters of wagon load EVDs

EVD based on	μ (kN)	σ (kN)	CoV (%)
1 wagon	1106	16.9	1.53
3 wagons	3119 (= 3@1040)	36.4	1.17
5 wagons	5090 (= 5@1018)	49.5	0.97
10 wagons	10,030 (= 10@1003)	91.9	0.92

Model uncertainty in the LL considers the uncertainty associated with the definition of load itself and the uncertainty that comes from transforming the load to a load effect. The CoV associated with a road LL model, V_{lf}, is taken as 10%, 15% or 20%, considering the loading uncertainty to be small, medium or large, respectively (DRD 2004). In the case of the railway structure analysed, it can be seen from Table 9.9 that the level of uncertainty associated with the load in the extreme is very low (CoV <1%) when the EVD is based on 10 wagons. The uncertainty associated with the load model V_{lf} is therefore taken as 'small', that is, 10%. A lower value could reasonably be argued on the basis of the low uncertainty associated with train loading relative to road traffic.

In the probabilistic assessment, the dynamic factor is modelled as per Equation 9.3 (DRD 2004). The dynamic increment, ε, employed in the deterministic assessment, is assumed to represent the 98% fractile value. For global effects, modelling the dynamic increment with mean and standard deviation equal to 0.03 and 0.03, respectively, gives a 98% value of 0.09 and a dynamic amplification factor $K_s = 1 + 0.09 = 1.09$, which corresponds well to the value employed in the deterministic assessment.

Incorporation of additional loads

For the elements and joints considered, the critical deterministic utilisation ratio was produced by a load combination rule of the form 1.0DL + 1.0SDL + 1.0LL + 0.4Brake + 0.6Wind + 0.6Temperature + 0.4SideImpact (O'Connor et al. 2009). In the probabilistic assessment, only the loads due to DL, SDL and LL are modelled stochastically. To allow for the effect of the additional loads (i.e. brake, wind, temperature and side impact) in the probabilistic assessment, a simple factored addition of their effects according to the load–combination rule, is performed. This stochastic–deterministic combination is only considered possible as the contribution of the deterministically modelled loads to the critical utilisation ratio is small in all cases (<10%). Incorporation of their effects in this manner is shown to be conservative. Were their contribution to the utilisation to be of significance, then they would have been modelled stochastically to reduce this conservatism. However, for the case under consideration, the adopted approach provides a practical solution.

9.4.3 Results of probabilistic assessment

The probability of failure, P_f, is found using a FORM analysis. The result is a computed probability of failure, $P_f = 7.13 \times 10^{-9}$ or $\beta_{FORM} = 5.67$. As the computed β exceeds the target level of 4.75, the structure has adequate safety at the considered limit state.

Importance factors

Table 9.10 presents the importance factors for this example.

Table 9.10 Importance factors

Variable	Importance (%)
Steel strength	70.2
Model uncertainty, traffic load	24.6
Dynamic increment	3.1
Dead load	1.6
Others	0.5

It is apparent that the strength of the steel and the LL model uncertainty with a cumulative importance of more than 94% are dominant. The other variables, for example, dynamic amplification, DL, SDL and so on, are of minor influence only.

Sensitivity analysis

The results of the sensitivity analysis are presented in Table 9.11. It is demonstrated that a +10% change in the mean strength of the steel, i.e. increase from 332.0 to 365.2 N/mm², increases the computed β by 1.38, from 5.67 to 7.05. As for previous examples, increasing the standard deviation reduces β. In this case, increasing the standard deviation by 10%, from 28.9 to 31.8 N/mm², changes the computed β by –0.43, from 5.67 to 5.24.

Values of the variables at the design point

As highlighted previously, reviewing the values of the variables at the β point is important to understand the extent to which certain variables have moved from their mean value into the tail of the distribution as opposed to those who have moved little, if at all, from the mean. Table 9.12 presents these values.

As expected, it is apparent from Table 9.12 that the variable that has moved most from its mean is the steel strength. At the β point, this variable has a value of 213.2 N/mm² compared to its mean value of 332 N/mm².

In this example, the safety of the element U_7 at the critical limit state has been demonstrated via a probabilistic assessment. At the outset, a number of critical locations were identified, but results for element U_7 only are presented. In these examples, the probabilistic assessment results demonstrate that the structural safety is in excess of the minimum acceptable level prescribed by the code. Of course, there will be cases where sufficient capacity cannot be shown. In these instances, the probabilistic models can be used to derive optimal repair strategies (O'Connor et al. 2009).

Table 9.11 Sensitivity analysis

Variable	Name	Distribution	$\Delta\beta$ for a +10% change in mean	$\Delta\beta$ for a +10% change in CoV
Steel strength	f_y	Lognormal	1.38	−0.43
Model uncertainty, traffic load	ModUnc	Normal	−0.49	−0.12
Dynamic increment	ε	Normal	−0.03	−0.01
Dead load	DL	Normal	−0.19	−0.01
Superimposed dead load	SDL	Normal	−0.04	0.0

Note: Change of β for a 10% change in the value of the mean value and CoV of the modelled parameters.

Table 9.12 Values of variables at β point

Variable	Name	Distribution	Mean, μ	Standard deviation, σ	Value of variable at β point, x_i^*
Steel strength	f_y	Lognormal	332 N/mm²	28.9 N/mm²	213.2 N/mm²
Model uncertainty, traffic load	ModUnc	Normal	1.0	0.1	1.24
Dynamic increment	ε	Normal	0.03	0.03	0.053
Dead load	DL	Normal	1.0	0.071	1.04
Superimposed dead load	SDL	Normal	1.0	0.112	1.03

9.5 CONCLUSION

The general deterministic approach to assessment of existing bridges using standard general codes is quick and efficient. However, it is conservative and unevenly so – some bridge types and some load effects will be assessed more conservatively than others. When there are problems with load-carrying capacity, this is costly due to unnecessary strengthening of bridges. The individual probabilistic approach described here is based on the concept that a bridge does not necessarily have to fulfil the specific requirements of a general code, as long as the overall level of safety defined by the code is satisfied. The individual approach is able to cut or reduce the strengthening or rehabilitation cost without compromising on the level of safety.

References

AASHTO (2010) *AASHTO LRFD Bridge Design Specifications*, American Association of State Highway and Transportation Officials, Washington.

AASHTO (2012) *AASHTO LFRD Bridge Design Specifications*, American Association of State Highway and Transportation Officials, Washington.

Andrade, C., Alonso, C. and Molina, F.J. (1993) 'Cover cracking as a function of bar corrosion: Part I – Experimental test', *Materials and Structures/Materiaux et Constructions*, **26**(162), 453–464.

Andrade, C., Diez, J.M. and Alonso, C. (1997) 'Mathematical modeling of a concrete surface 'skin effect' on diffusion in chloride contaminated media', *Advanced Cement Based Materials*, **6**(2), 39–44.

Ang, A.H.S. and Tang, W.H. (1984) *Probability Concepts in Engineering Planning and Design, Vol. 2, Decision, Risk and Reliability*, Wiley, New York, USA.

Armer, G.S.T. (1968) Correspondence, *Concrete*, **2**, August, 319–320.

BA42/96 (2003) *Design Manual for Roads and Bridges, Volume 1, Section 3, Part 12, Design of Integral Bridges*, (incorporating Amendment No. 1), Department for Transport, London. Available from http://www.dft.gov.uk/ha/standards/dmrb/.

Bakht, B. and Jaeger, L.G. (1997) 'Evaluation by proof testing of a T-beam bridge without drawings', *The Structural Engineer*, **75**(19), 339–344.

Bakht, B., Jaeger, L.G., Cheung, M.S. and Mufti, A.A. (1981) 'The state of the art in analysis of cellular and voided slab bridges', *Canadian Journal of Civil Engineering*, **8**, 376–391.

Bazant, Z.P. (1979) 'Physical model for steel corrosion in concrete sea structures – theory', *Journal of Structural Division, ASCE*, **105**(6), 1137–1153.

BD57/01 (2001) *Design Manual for Roads and Bridges, Volume 1, Section 3, Part 7, Design for Durability*, Department for Transport, London. Available from http://www.dft.gov.uk/ha/standards/dmrb.

BD81/02 (2002) *Design Manual for Roads and Bridges, Volume 3, Section 4, Part 20, Use of Compressive Membrane Action in Bridge Decks*, Department for Transport, London. Available from http://www.dft.gov.uk/ha/standards/dmrb.

Brady, S.P. and OBrien, E.J. (2006) 'The effect of vehicle velocity on the dynamic amplification of two vehicles crossing a simply supported bridge', *Journal of Bridge Engineering, ASCE*, **11**(2), 250–256.

Brady, S.P., OBrien, E.J. and Žnidarič, A. (2006) 'The effect of vehicle velocity on the dynamic amplification of a vehicle crossing a simply supported bridge', *Journal of Bridge Engineering, ASCE*, **11**(2), 241–249.

Caprani, C.C., González, A., Rattigan, P.H. and OBrien, E.J. (2012) 'Assessment dynamic ratio for traffic loading on highway bridges', *Structure and Infrastructure Engineering*, **8**(3), 295–304.

Caprani, C.C. and OBrien, E.J. (2010) 'The use of predictive likelihood to estimate the distribution of extreme bridge traffic load effect', *Structural Safety*, **32**(2), 138–144.

Caquot, A. and Kersiel, J. (1948) *Tables for the Calculation of Passive Pressure, Active Pressure and Bearing Capacity of Foundations* (translated from French by M.A. Bec), Gauthier-Villars, Paris.

Castillo, E. (1988) *Extreme Value Theory in Engineering*, Academic Press, New York.

Cebon, D. (1999) *Handbook of Vehicle-Road Interaction*, Swets & Zeitlinger, Lisse, The Netherlands.

Chernin, L. and Val, D. (2011) 'Prediction of corrosion-induced cover cracking in reinforced concrete structures', *Construction and Building Materials*, **25**, 1854–1869.

Clark, L.A. and Sugie, I. (1997) 'Serviceability limit state aspects of continuous bridges using precast concrete beams', *The Structural Engineer*, **75**(11), 185–190.

COST345 (2004) *Procedures Required for the Assessment of Highway Structures, Numerical Techniques for Safety and Serviceability Assessment*, Report of Working Groups 4 and 6. Cooperation in the Field of Scientific and Technical Research, Brussels. Available from http://cost345.zag.si (accessed 3 July 2014).

Cremona, C. (2001) 'Optimal extrapolation of traffic load effects', *Structural Safety*, **23**(1), 31–46.

Das, P. (1997) *Safety of Bridges*, Thomas Telford, London.

Dawe, P. (2003) *Research Perspectives: Traffic Loading on Highway Bridges,* Thomas Telford, London.

Denton, S.R. and Burgoyne, C.J. (1996) 'The assessment of reinforced concrete slabs', *The Structural Engineer*, **74**(9), 147–152.

Ditlevsen, O. (1994) 'Traffic loads on large bridges modeled as white noise fields', *Journal of Engineering Mechanics*, **120**(4), 681–694.

Ditlevsen, O. and Madsen, H.O. (1994) 'Stochastic vehicle queue load model for large bridges', *Journal of Engineering Mechanics*, **120**(9), 1829–1847.

Dobry, R. and Gazetas, G. (1986) 'Dynamic response of arbitrarily shaped foundations', *Journal of Geotechnical Engineering, ASCE*, **112**(2), 109–135.

DRD (2004) *Report 291, Reliability-Based Classification of the Load Carrying Capacity of Existing Bridges,* Road Directorate, Ministry of Transport, Denmark.

El Maaddawy, T. and Soudki, K. (2007) 'A model for prediction of time from corrosion initiation to corrosion cracking', *Cement and Concrete Composites*, **29**, 168–175.

Ellingwood, B., MacGregor, J.G., Galambos, T.V. and Cornell, A.C. (1982) 'Probability based load criteria: Load factors and load combinations', *Journal of the Structural Division, ASCE*, **108**(5), 978–997.

EN 1990 (2002) *Eurocode – Basis of Structural Design*. European Committee for Standardisation, Brussels.

EN 1991-1 (2002) *Eurocode 1: Actions on Structures, Part 1: General Actions*, European Committee for Standardisation, Brussels.

EN 1991-2 (2003) *Eurocode 1: Actions on Structures, Part 2: Traffic Loads on Bridges*, European Committee for Standardisation, Brussels.

EN 1997 (2004) *Eurocode 7: Geotechnical Design, Part 1: General Rules, EN 1997–1:2004*, European Committee for Standardisation, Brussels.

Engelund, S. and Sorensen, J.D. (1998) 'A probabilistic model for chloride-ingress and initiation of corrosion in reinforced concrete structures', *Structural Safety*, **20**(1), 69–89.

England, G.L., Tsang, N.C.M. and Bush, D.I. (2000) *Integral Bridges – A Fundamental Approach to the Time Temperature Loading Problem,* Thomas Telford, London.

Enright, B. and OBrien, E.J. (2013) 'Monte Carlo simulation of extreme traffic loading on short and medium span bridges', *Structure and Infrastructure Engineering*, **9**(12), 1267–1282.

Enright, B., OBrien, E.J. and Leahy, C. (2014) 'The importance of permit trucks in critical bridge loading events', submitted for publication.

Fib (2012) *Bulletin 65: Model Code 2010*, Volume 1. Fédération Internationale du Beton.

Frýba, L. (1971) *Vibration of Solids and Structures under Moving Loads*. Noordhoff International Publishing, Groningen, The Netherlands.

Galambos, T.V., Ellingwood, B., MacGregor, J.G. and Cornell, A.C. (1982) 'Probability based load criteria: Assessment of current design practice', *Journal of the Structural Division, ASCE*, **108**(5), 959–977.

Ghali, A., Neville, A.M. and Brown, T.G. (2009) *Structural Analysis: A Unified Classical and Matrix Approach,* 6th edn, CRC Press.

Ghosn, M. and Moses, F. (1986) 'Reliability calibration of bridge design code', *Journal of Structural Engineering, ASCE*, **112**(4), 745–763.

González, A., Cantero, D. and OBrien, E.J. (2011) 'Dynamic increment in shear load effect due to heavy traffic crossing a highway bridge', *Computers and Structures*, **89**(23), 2261–2272.

González, A., Rowley, C. and OBrien, E.J. (2008) 'A general solution to the identification of moving vehicle forces on a bridge', *International Journal for Numerical Methods in Engineering*, 75(3), 335–354.

Haldar, A. and Mahadevan, S. (2000) *Probability, Reliability and Statistical Methods in Engineering Design*, John Wiley & Sons, New York; Chichester.

Hambly, E.C. (1991) *Bridge Deck Behaviour*, 2nd edn, E&FN Spon, London.

Harney, R. (2012) *Developing Finite Element Solution for Shear Forces in Bridges*, ME Dissertation, University College Dublin, Ireland.

Hasofer, A.M. and Lind, N.C. (1974) 'Exact and invariant second moment code format', *Journal of the Engineering Mechanics Division, ASCE*, **100**(1), 111–121.

Hewson, N.R. (2003) *Prestressed Concrete Bridges: Design and Construction*, Thomas Telford, London.

Heywood, R., Roberts, W. and Boully, G. (2001) 'Dynamic loading of bridges' *Transportation Research Record* 1770: 58–66, Transportation Research Board, Washington, D.C.

Hwang, E.S. and Nowak, A.S. (1991) 'Simulation of dynamic loads on bridges', *Journal of Structural Engineering*, **119**(6), 853–867.

Indraratna, B., Vinod, J.S. and Lackenby, J. (2009) 'Influence of particle breakage on the resilient modulus of railway ballast', *Geotechnique* **59**(7), 643–646.

ISO/CD 13822 (2010) *Basis for Design of Structures – Assessment of Existing Structures*. International Organisation for Standardisation.

JCSS (2000a) *Probabilistic Model Code I – Basis of Design*. Joint Committee of Structural Safety.

JCSS (2000b) *Probabilistic Model Code III – Resistance Models*. Joint Committee of Structural Safety.

JCSS (2006) *Probabilistic Model Code, Section 3.7: Soil Properties*. Joint Committee of Structural Safety.

Karimi, A., Ramachandran, K. and Buenfeld, N.R. (2005) 'Probabilistic analysis of reinforcement corrosion with spatial variability', *ICOSSAR Conference*, Rotterdam, 679–686.

Kenshel, O. (2009) Influence of Spatial Variability on Whole Life Management of Reinforced Concrete Bridges, Ph.D. thesis, Trinity College Dublin, Dublin, Ireland.

Keogh, D.L. and OBrien, E.J. (1996) 'Recommendations on the use of a 3-D grillage model for bridge deck analysis', *Structural Engineering Review*, **8**(4), 357–366.

Kirkegaard, P.H., Nielsen, S.R.K. and Enevoldsen, I. (1997) *Heavy Vehicles on Minor Highway Bridges – Dynamic Modelling of Vehicles and Bridges*, Structural reliability theory paper no. 171, Aalborg University, ISSN 1395-7953-R9721.

Lee, D.J. (1994) *Bridge Bearings and Expansion Joints*, 2nd edn, E&FN Spon, London.

Lehane, B. (1999) 'Predicting the restraint to integral bridge deck expansion', in *Proceedings of 12th European Conference on Soil Mechanics and Geotechnical Engineering*, June, Balkema, Rotterdam.

Lehane, B.M. (2011) 'Lateral soil stiffness adjacent to deep integral bridge abutments', *Geotechnique*, **61**(7), 593–603.

Lehane, B., Keogh, D.L. and OBrien, E.J. (1999) 'Simplified elastic model for restraining effects of back-fill soil on integral bridges', *Computers and Structures*, **73**(1–5), 303–313.

Leonhardt, F. (1983) *Bridges, Aesthetics and Design*, Deutsche Verlags-Anstalt, Stuttgart.

Li, Y., Vrouwenvelder, T., Wijnants, G.H. and Walraven, J. (2004) 'Spatial variability of concrete deterioration and repair strategies', *Structural Concrete*, **5**(3), 121–129.

Liu, Y. and Weyers, R.E. (1998) 'Modeling the time-to-corrosion cracking in chloride contaminated reinforced concrete structures', *ACI Materials Journal*, **95**(6), 675–681.

Madsen, H.O., Krenk, S. and Lind, N.C. (1986) *Methods of Structural Safety*, Prentice Hall, New Jersey.

McNulty, P. and OBrien, E.J. (2003) 'Testing of bridge weigh-in-motion system in sub-arctic climate', *Journal of Testing and Evaluation*, **31**(6), 1–10.

Melchers, R.E. (1999) *Structural Reliability; Analysis and Prediction*, 2nd edn, John Wiley & Sons, New York; Chichester.

Minervino, C., Sivakumar, B., Moses, F., Mertz, D. and Edberg, W. (2004) 'New AASHTO guide manual for load and resistance factor rating of highway bridges', *Journal of Bridge Engineering, ASCE*, **9**(1), 43–54.

Mirza, S.A. and MacGregor, J.G. (1979) 'Variability of mechanical properties of reinforcing bars', *Journal of Structural Division, ASCE*, **105**(ST5), 921–937.

Mullard, J.A. and Stewart, M.G. (2011) 'Corrosion-induced cover cracking: New test data and predictive models', *ACI Structural Journal*, **108**(1), 71–79.

Nielson, B.G. and DesRoches, R. (2007) 'Seismic fragility methodology for highway bridges using a component level approach', *Earthquake Engineering and Structural Dynamics*, **36**, 823–839.

NKB (1978) *Report No. 36 Guidelines for Loading and Safety Regulations for Structural Design*. Nordisk Komité for Bygningsbestemmelser.

Nowak, A. and Szerszen, M.M. (1998) 'Bridge load and resistance models', *Engineering Structures*, **20**(11), 985–990.

OBrien, E.J., Cantero, D., Enright, B. and González, A. (2010) 'Characteristic dynamic increment for extreme traffic loading events on short and medium span highway bridges', *Engineering Structures*, **32**(12), 3287–3835.

OBrien, E.J. and Enright, B. (2011) 'Modeling same-direction two-lane traffic for bridge loading', *Structural Safety*, **33**(4–5), 296–304.

OBrien, E.J. and Enright, B. (2013) 'Using weigh-in-motion data to determine aggressiveness of traffic for bridge loading', *Journal of Bridge Engineering, ASCE*, **18**(3), 232–239.

OBrien, E.J., González, A., Dowling, J. and Žnidarič, A. (2013) 'Direct measurement of dynamics in road bridges using a bridge weigh-in-motion system', *The Baltic Journal of Road and Bridge Engineering*, **8**(4), 263–270.

OBrien, E.J. and Keogh, D.L. (1998) 'Upstand finite element analysis of slab bridges', *Computers and Structures*, **69**, 671–683.

OBrien, E.J., Rattigan, P., González, A., Dowling, J. and Žnidarič, A. (2009) 'Characteristic dynamic traffic load effects in bridges', *Engineering Structures*, **31**(7), 1607–1612.

OBrien, E.J., Žnidarič, A. and Dempsey, A.T. (1999) 'Comparison of two independently developed bridge weigh-in-motion systems', *Heavy Vehicle Systems, International Journal of Vehicle Design*, **6**(1/4), 147–161.

OBrien, S.G. (1997) *The Analysis of Shear Forces in Slab Bridge Decks*, MSc Thesis, University of Dublin, Trinity College, Dublin.

OBrien, S.G., OBrien, E.J. and Keogh, D.L. (1997) 'The calculation of shear force in prestressed concrete bridge slabs', in *The Concrete Way to Development*, FIP Symposium, Johannesburg, South Africa, 233–237.

O'Connor, A.J. and Eichinger, E. (2007) 'Site-specific traffic load modelling for bridge assessment', *Bridge Engineering, Proceedings of the Institution of Civil Engineers*, **160**(BE4), 185–194.

O'Connor, A. and Enevoldsen, I. (2008) 'Probability based assessment of an existing prestressed post-tensioned concrete bridge', *Engineering Structures*, **30**(2008), 1408–1416.

O'Connor, A.J., Jacob, B., OBrien, E.J. and Prat, M. (2001) 'Report of current studies performed on normal load model of EC1 – traffic loads on bridges', **5**(4), *RFGC, Hermes Science Publications*, 411–434.

O'Connor, A. and Kenshel, O. (2012) 'Experimental evaluation of the scale of fluctuation for spatial variability modelling of chloride induced reinforced concrete corrosion', *Journal of Bridge Engineering, ASCE*, **18**(1), 3–15.

O'Connor, A.J. and OBrien, E.J. (2005) 'Mathematical traffic load modelling and factors influencing the accuracy of predicted extremes', *Canadian Journal of Civil Engineering*, **32**(1), 270–278.

O'Connor, A., Pedersen, C., Gustavsson, L. and Enevoldsen, I. (2009) 'Probability based assessment and optimised maintenance planning for a large riveted truss railway bridge', *Structural Engineering International*, **19**(4), 375–383.

OHBDC (1992) *Ontario Highway Bridge Design Code*, Ministry of Transportation of Ontario, Downsview, Ontario, Canada.

PIARC (1999) *Reliability-Based Assessment of Highway Bridges*. Technical Committee 11 Bridges and Other Structures. World Road Association.

Reale, T. and O'Connor, A. (2012) 'A review and comparative analysis of corrosion induced time to first crack models', *Journal of Construction and Building Materials*, **36**(2012), 475–483.

Richardson, J., Jones, S., Brown, A., OBrien, E.J. and Hajializadeh, D. (2014) 'On the use of bridge weigh-in-motion for overweight truck enforcement', *International Journal of Heavy Vehicle Systems*, **21**(2), 83–104.

SAMCO (2006) *Final Report – F08a Guideline for the Assessment of Existing Structures*, Structural Assessment Monitoring and Control, 2006. Available from http://www.samco.org/ (accessed 3 July 2014).

Schneider, J. (1997) *Introduction to Safety and Reliability of Structures*. Structural Engineering Documents, IABSE, Zurich.

Sivakumar, B., Moses, F., Fu, G. and Ghosn, M. (2007) *Legal truck loads and AASHTO Legal Loads for Posting*, NCHRP Report 575.

Springman, S.M., Norrish, A.R.M. and Ng, C.W.W. (1996) *Cyclic Loading of Sand Behind Integral Bridge Abutments*, TRL Report 146, UK Highways Agency, London.

Stewart, M.G. (2009) 'Mechanical behaviour of pitting corrosion of flexural and shear reinforcement and its effect on structural reliability of corroding RC beams', *Structural Safety*, 31(1), 19–30.

Stewart, M.G. and Mullard, J.A. (2007) 'Spatial time-dependent reliability analysis of corrosion damage and the timing of first repair for RC structures', *Engineering Structures*, 29(7), 1457–1464.

Stewart, M.G. and Rosowsky, D.V. (1998) 'Structural safety and serviceability of concrete bridges subject to corrosion', *Journal of Infrastructure Systems*, 4(4), 146–155.

Sykora, M. and Holicky, M. (2011) 'Target reliability levels for assessment of existing structures', *International Conference on Application of Statistics and Probability in Civil Engineering – ICASP 2011*, Zurich.

Taylor, H.P.J., Clark, L.A. and Banks, C.C. (1990) 'The Y-beam: A replacement for the M-beam in beam and slab bridge decks', *The Structural Engineer*, 68(23), 459–465.

Tilly, G. (2007) 'The durability of repaired concrete structures', Proceedings of the International Association for Bridge and Structural Engineering Symposium (IABSE), Weimar, Germany.

Timoshenko, S.P. and Goodier, J.N. (1970) *Theory of Elasticity*, 3rd edn, McGraw-Hill, New York.

Timoshenko, S.P. and Woinowsky-Krieger, S. (1970) *Theory of Plates and Shells*, 2nd edn, McGraw-Hill, Singapore.

Toft-Christensen, P. and Baker, M.J. (1982) *Structural Reliability Theory and its Applications*, Springer-Verlag.

Tomlinson, M.J. (1994) *Pile Design and Construction Practice*, 4th edn, E&FN Spon, London.

Troitsky, M.S. (1967) *Orthotropic Bridges: Theory and Design*, James F. Lincoln Arc Welding Foundation, Cleveland, Ohio.

Tutti, K. (1982) *Corrosion of Steel in Concrete*, Swedish Cement and Concrete Research Institute, Research Report FO 4, Stockholm, Sweden.

Val, D.V. and Melchers, R.E. (1997) 'Reliability of deteriorating RC slab bridges', *Journal of Structural Engineering, ASCE*, 123(12), 1638–1644.

Vu, K.A.T. and Stewart, M.G. (2005) 'Predicting the likelihood and extent of reinforced concrete corrosion-induced cracking', *Journal of Structural Engineering*, 131(11), 1681–1689.

Vu, K.A.T., Stewart, M.G. and Mullard, J.A. (2005) 'Corrosion-induced cracking: Experimental data and predictive models', *ACI Structural Journal*, 102(5), 719–726.

Weyers, R.E., Chamberlin, W.P., Hoffman, P. and Cady, P.D. (1991) *Concrete bridge protection and rehabilitation: Chemical and physical techniques*, Task 1 Interim report (SHRP-87-C-103), Strategic Highway Research Program. Virginia Polytechnic Institute and State University. Blacksburg.

Wiśniewski, D.F., Cruz, P.J.S., Henriques, A.A.R. and Simões, R.A.D. (2012). 'Probabilistic models for mechanical properties of concrete, reinforcing steel and prestressing steel'. *Structure and Infrastructure Engineering*, 8(2), 111–123.

Wood, R.H. (1968) 'The reinforcement of slabs in accordance with a pre-determined field of moments', *Concrete*, 2, 69–76.

Zienkiewicz, O.C. and Cheung, Y.K. (1964) 'The finite element method for analysis of elastic isotropic and orthotropic slabs', *Proceedings of the Institution of Civil Engineers*, 28, 471–488.

Zienkiewicz, O.C., Taylor, R.L. and Zhu, J.Z. (2013) *The Finite Element Method*, 7th edn, Elsevier.

Žnidarič, A. and Lavrič, I. (2010) 'Applications of B-WIM technology to bridge assessment', *Bridge Maintenance, Safety, Management and Life-Cycle Optimization*, Eds. D.M. Frangopol, R. Sause and C.S. Kusko, Philadelphia, Taylor & Francis, 1001–1008.

Appendix A: Stiffness of structural members and associated bending moment diagrams

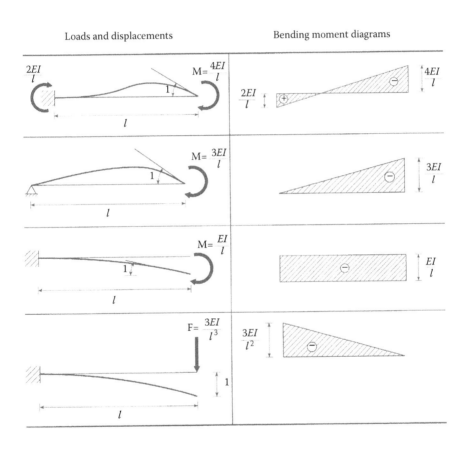

Loads and displacements	Bending moment diagrams

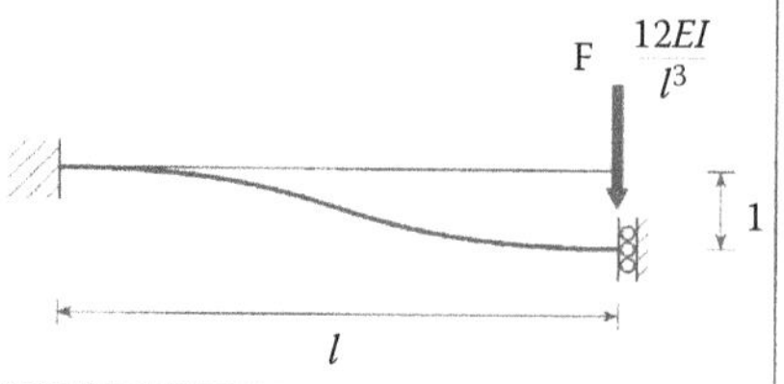

F $\dfrac{12EI}{l^3}$ 1 l

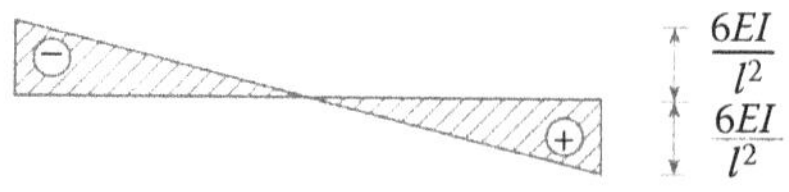

$\dfrac{6EI}{l^2}$
$\dfrac{6EI}{l^2}$

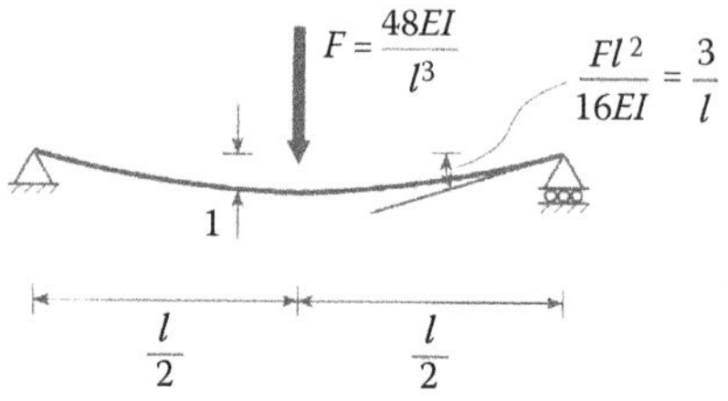

$F = \dfrac{48EI}{l^3}$ $\dfrac{Fl^2}{16EI} = \dfrac{3}{l}$ 1 $\dfrac{l}{2}$ $\dfrac{l}{2}$

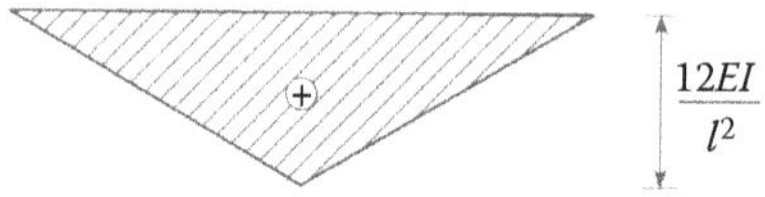

$\dfrac{12EI}{l^2}$

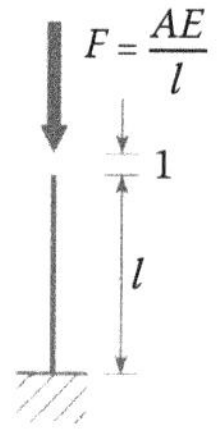

$F = \dfrac{AE}{l}$ 1 l

No bending moment

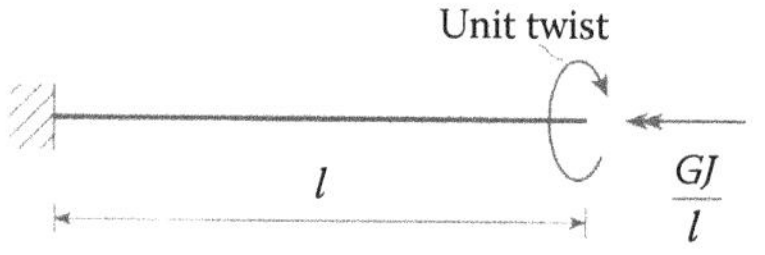

Unit twist l $\dfrac{GJ}{l}$

No bending moment
but constant torsion of GJ/l

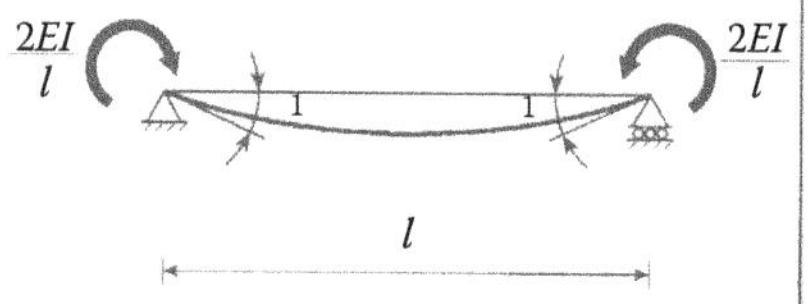

$\dfrac{2EI}{l}$ 1   1 $\dfrac{2EI}{l}$ l

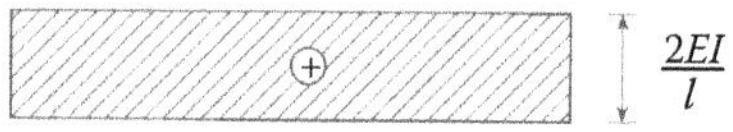

$\dfrac{2EI}{l}$

Appendix B: Location of centroid of a section

The centroid, $\bar{y}$, of any section can be found from the coordinates of the perimeter points using the formula

$$\bar{y} = \frac{\sum_{i=1}^{n}(x_i - x_{i+1})\left(y_i^2 + y_i y_{i+1} + y_{i+1}^2\right)}{3\sum_{i=1}^{n}(x_i - x_{i+1})(y_i + y_{i+1})} \tag{B.1}$$

where x_i and y_i are the coordinates of point i and n is the number of coordinate points. For the purposes of this calculation, point $n + 1$ is defined as equal to point 1. For the section of Figure 6.6, the coordinates are taken from the figure, starting at the top left corner and specifying only half the section (which will have the same centroid as the full section). The terms of Equation B.1 are given in Table B.1 where the 'Top' and 'Bottom' columns refer to the numerator and denominator, respectively, of the fraction specified in the equation.

The y coordinate of the centroid is then

$$\bar{y} = \frac{-20.64 \times 10^9}{-32.40 \times 10^6} = 637 \text{ mm} \tag{B.2}$$

The same answer can be found by dividing the section into rectangles and triangles and summing moments of area about any common point.

Table B.1 Evaluation of Equation B.1

x_i	y_i	$(x_i - x_{i+1})$	$y_i^2 + y_iy_{i+1} + y_{i+1}^2$	$y_i + y_{i+1}$	Top	Bottom
0	1200	−5500	4,320,000	2400	-23.76×10^9	-39.60×10^6
5500	1200	0	1,440,000	1200	0	0
5500	0	4000	0	0	0	0
1500	0	300	640,000	800	0.19×10^9	0.72×10^6
1200	800	1200	2,440,000	1800	2.93×10^9	6.48×10^6
0	1000	0	3,640,000	2200	0	0
0	1200	0	4,320,000	2400	0	0
				Sum =	-20.64×10^9	-32.40×10^6

Appendix C: Derivation of shear area for grillage member representing cell with flange and web distortion

The transverse shear force halfway across the cell will be distributed between the flanges in proportion to their stiffness. Hence, the shear force in the top flange will be

$$\frac{V(i_t)}{(i_t + i_b)}$$

where V is the total shear force and i_t and i_b are the second moments of area per unit breadth of the top and bottom flanges, respectively. This force is illustrated in Figure C.1 for a segment of cell between points of contraflexure. Hence, the total moment at the top of the web is

$$M = 2\frac{Vi_t}{(i_t + i_b)}\frac{l}{2} = \frac{Vli_t}{(i_t + i_b)}$$

The rotation of the web due to this moment is

$$\theta = \frac{M\left(\dfrac{h}{2}\right)}{3Ei_w} = \frac{Vli_t h}{6(i_t + i_b)Ei_w}$$

where h is the bridge depth (centre to centre of flanges) and i_w is the web second moment of area per unit breadth. The total deflection in the top flange results from this rotation plus bending in the flange itself:

$$\delta_t = 2\left(\frac{l}{2}\right)\theta + 2\left(\frac{Vi_t}{(i_t + i_b)}\right)\left(\frac{l}{2}\right)^3\left(\frac{1}{3Ei_t}\right)$$

$$= \frac{Vl^2 h i_t}{6Ei_w(i_t + i_b)} + \frac{Vl^3}{12E(i_t + i_b)}$$

Similarly, the deflection in the bottom flange can be shown to be

$$\delta_b = \frac{Vl^2 h i_b}{6Ei_w(i_t + i_b)} + \frac{Vl^3}{12E(i_t + i_b)}$$

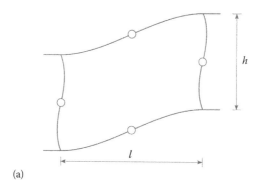

(a)

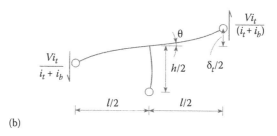

(b)

Figure C.1 Cell with flange and web distortion: (a) assumed distortion; (b) segment of cell between points of contraflexure.

The mean deflection is

$$\delta_{av} = \frac{Vl^2h(i_t + i_b) + Vl^3 i_w}{12Ei_w(i_t + i_b)}$$

Equating this to the shear deformation in a grillage member gives

$$\frac{Vl^2h(i_t + i_b) + Vl^3 i_w}{12Ei_w(i_t + i_b)} = \frac{Vl}{Ga_s}$$

$$\Rightarrow a_s = \left(\frac{l}{G}\right)\left(\frac{12Ei_w(i_t + i_b)}{l^2h(i_t + i_b) + l^3 i_w}\right)$$

$$= \left(\frac{E}{G}\right)\left(\frac{12i_w(i_t + i_b)}{lh(i_t + i_b) + l^2 i_w}\right)$$

If the second moments of area per unit breadth are expressed in terms of the flange and web depths ($i_t = d_t^3/12$, etc.), this becomes Equation 6.7:

$$a_s = \left(\frac{E}{G}\right)\left(\frac{d_w^3\left(d_t^3 + d_b^3\right)}{lh\left(d_t^3 + d_b^3\right) + l^2 d_w^3}\right)$$

Index

9 780367 869397